CHANGING WOMEN'S LIVES AND WORK

Changing Women's Lives and Work

An analysis of the impacts of eight microenterprise projects

LUCY E. CREEVEY

IT PUBLICATIONS 1996

Published by Intermediate Technology Publications
103–105 Southampton Row, London WC1B 4HH, UK

© UNIFEM 1996

ISBN 1 85339 319 3

A CIP catalogue record for this book is available from the
British Library

Typeset by Dorwyn Ltd, Rowlands Castle, Hants, UK
Printed by BPC Wheatons, Exeter, UK

Contents

PREFACE	ix
ONE: Introduction	1
Framework for analysis	5
Sponsoring agencies: UNIFEM, ITDG and ATI	6
United Nations Development Fund for Women	7
Intermediate Technology Development Group	8
Appropriate Technology International	9
Strategy choice in eight SME projects	11
Ranking the countries	13
Research methodology	16
The questionnaires	16
Supporting research	19
Guide to the discussion	21
Setting the bias out front	21
Conclusion	23
TWO: Food Processing in Peru	24
Setting the context	24
Women in the Peruvian economy	26
The Peru Food Processing Project	29
Facing the field	32
The women's report	35
Conclusion	42
Annex	45
THREE: Honduras – Cashew Nut and Fruit Production	46
The state and the nation	46
Women in Honduras	47
The Honduran Cashew Project 1981–8	50
Events and challenges	52
The women's record	56
Conclusion	61
Annex	66

FOUR: Guatemala – Wool Production and Processing Project   67
   Economy and politics   67
   Guatemalan women   69
   The Wool Production and Processing Project   71
      Sheep and Wool Production   72
      Wool Processing and Weaving   73
      Commercial Enterprise   73
   Overcoming the obstacles   74
   Men and women react   78
   The bottom line   86
   Annex   90

FIVE: Bangladesh – Surjosnato Coconut Products   91
   Poverty and hope   91
   Growing forces of conservatism: women in Bangladesh   93
   The Surjosnato Coconut Project   94
   Dealing with the impossible   97
   Women speak out   100
   Balancing the results   106
   Annex   109

SIX: India – The Sericulture Project   110
   India and development   110
   Indian women   111
   The Women's Sericulture Project   114
   Establishing the reality   118
   Impacts on women   120
   Assessing the record   127
   Annex   131

SEVEN: Thailand – Venture Capital and the Pickled Ginger Group   132
   The development miracle   132
   Quiet power: Thai women   133
   The Joint Venture Capital Project in Northern Thailand   135
   Venture capital in action   138
   Impacts and changes in individual lives   139
   Expectations and reality   146
   Annex   149

EIGHT: Ghana – Shea Butter Processing   150
   Structural adjustment and development   150
   Women in the marketplace   153
   The Shea Butter Processing Project in Ghana   155
   The project in the field   158

Many faces, many experiences	161
The reality of the Shea Butter Project	167
Annex	170

NINE: Tanzania – Food Processing — 171
Climbing up the development ladder	171
Poverty and prospects: Tanzanian women	175
The Tanzania Food Processing Programme	178
The dream and the reality	182
Women's voices	182
Realizing the dream	189
Annex	195

TEN: A Comparison of the Impacts of Different Strategies ... — 196
Basis of comparison	196
Comparative data analysis	200
Findings	201
Significance of quantitative analysis findings	208
Conclusion	209
Annex	220
BIBLIOGRAPHY	222

For my sons Kennedy and Ari

Preface

A MAJOR PART of the research for this book was conducted by Carolyn McCommon and Valerie Autissier. They visited all eight countries where the projects under consideration were located. Carolyn McCommon went to Peru, Honduras and Guatemala, while Valerie Autissier went to Bangladesh, India, Thailand, Ghana and Tanzania. In all countries, both researchers interviewed project staff and some participants, studied project documents and trained the local researchers in the administration of the individual and enterprise questionnaires, the general purpose and orientation of the study, and the role of the various researchers in the work. They also consulted staff members in the three international agencies (UNIFEM, the funding sponsor, with ATI and ITDG, collaborating sponsors) on project context, history and perceived results. At the end of their work, they submitted extensive reports on each case which were the principal sources for the case studies presented here.

The local researchers were: Marisela Benavides (Peru), Ana Ruth Zuniga Izzaguirre (Honduras), Luz Marina Delgado (Guatemala), Afsana Wahab (Bangladesh), Ila Varma (India), Rachitta Na Pataling (Thailand), Mary Abena Kyei (Ghana) and I.H. Mafwenga (Tanzania). They administered the individual questionnaires to samples of about eighty individuals in each country and questionnaires investigating the enterprise experience directly to a small sample of enterprise owners/managers. They kept field notes, answered numerous questions about the climatic, economic and political conditions faced by the projects, interviewed project staff and wrote lengthy observations and comments on the questionnaires as well. The completed questionnaires and their notes and responses were sent on to me in Storrs (Connecticut) by the end of September 1993.

The final report presented to UNIFEM (in association with both ITDG and ATI) in June 1994 was written by me as the principal investigator but it was based centrally on the work of these ten women and particularly on the observations of Carolyn McCommon and Valerie Autissier. The report was, indeed, a fully collaborative effort, although only I am responsible for the conclusions or interpretations presented in it. This book, which resulted from the report, was also prepared by me, completed in March 1995, and revised for publication in August 1995. I made extensive revisions in order to open up the discussion to bring in people who are not academics or planners but who are vitally concerned about women in developing countries. In these revisions, however, I did not alter the research findings but instead attempted to interpret them more clearly and meaningfully. I am again solely responsible for the results.

I am extremely grateful for the extensive, in-depth work of Valerie and Carolyn and the ten local researchers, without which this study would have had no substance. I am also very grateful to Valerie for her corrections and comments on the first draft I prepared, and to Carolyn who has answered countless questions and commented on all drafts of the report at great length. I would also like to thank the numerous people in the local agencies in the field who patiently responded to countless questions and even requests, once the surveys were done, for follow-up information.

Finally, each of the three international agencies helped in the preparation of this report, enduring frequent questions and providing countless detailed documents. I cannot acknowledge everyone who helped Valerie and Carolyn and me but I would like to mention a few. At ITDG, Freer Spreckley helped establish this research and supported Valerie's efforts throughout her work. Barry Axtell answered many queries and gave very helpful comments and Tris Bartlett read and commented on the final draft. At ATI, Jeanne Downing answered many questions and gave me very valuable comments on the questionnaires. Valeria Budinich answered repeated questions for Carolyn, while S.K. Gupta gave extensively of his time in clarifying my perception of the Thai project. Among several others, Jack Croucher and Carlos Lola also helped by providing insights into the experiences of different projects. The person to whom this study is most indebted, however, is Eric Hyman who provided extensive and detailed suggestions and commentaries on questionnaires and on the preliminary draft, and was always ready to provide needed information. He is not responsible for what is written but this study would have been much poorer without him.

At UNIFEM, many people helped in this study, providing information, criticizing drafts and suggesting different ways of considering or interpreting the material. Among these were Simone Buechler, who helped me with suggestions and documentation throughout the research, Marilyn Carr, who offered many useful ideas and interesting sources, especially at the beginning, and Joanne Sandler who read and made suggestions on all the drafts of the report and the book. The person who bore the full weight of this research, however, was Teckie Ghebre-Medhin, whose idea the study was in the first place. He kept in his hands the links to all the researchers and handled all the support for everyone. Repeatedly, he solved the problems which arose and helped everyone adjust to the new situations. Teckie was the mainstay of the study and the report and for this book. If it has a measure of success, this is due to his efforts.

In the early stages of writing this book, I profited greatly from a workshop held by UNIFEM in October 1994 where some of the major findings from the report were discussed. I can not thank all the people individually who took the time to come and make comments and suggestions but I am extremely appreciative of their efforts. This book is immeasurably improved as a result.

In addition, many people helped me in Storrs, not all of whom I can mention here. I am most grateful to Momar N'Diaye for his help in data analysis, Abdourahman Thiam and Theodoro Rivas for their help in coding and processing the data, and Natalia Brandler and Guadelupe Rodriquez-Streeter for their helpful insights on India and Guatemala. Jennifer Vincent has been invaluable in helping edit and correct the final draft of the report and the draft of the revised book manuscript. Jere Behrman gave me comments and suggestions on the comparative data analysis, and Ton de Wilde read and commented on the entire report. Donna May helped me with the production of both drafts of the report. I am deeply grateful to all of them.

LUCY E. CREEVEY
STORRS, CONNECTICUT
AUGUST 1995

CHAPTER ONE
Introduction

SUPPORTING SMALL OR medium-scale enterprise development has become popular among planners and policy-makers in poor countries everywhere. The advantage to this strategy is in capitalizing on the entrepreneurial instincts of people. In urban and periurban areas, where the poor and unemployed are clustered, micro, small and medium-scale enterprise projects support a wide variety of enterprises in all sectors including manufacturing, trade and service. Many types of rural enterprise may also be encompassed in this approach. Some examples are provided by non-traditional agricultural production (truck gardening, fruit growing or animal fattening) carried out with a direct connection to marketing possibilities rather than simply for subsistence.

The popularity of the small enterprise (SME) approach among American policy makers – with its full-cycle concern and its stress on the profit motive – is easy to understand (McKean and Binnendjik, 1988; Liedholm and Mead, 1987; Little, Mazumder and Page, 1987). However, the small enterprise strategy has gained credence throughout the world of planners concerned with effective development assistance, not just those advancing capitalism. In the continuing record of failures in development projects, small-scale enterprise projects have a better report (French, 1988; Grown and Sebstad, 1989:943; Buvinic, 1989:1051). For one thing, the small enterprise/microenterprise approach looks at the informal sector activities of the inhabitants of the poorest developing countries. It seeks to support the things which poor and untrained people can do. It makes available to them credit and inputs which they often can not get (De Wilde and Schreurs, 1991:Introduction). Supporting SMEs is also consistent with the current international wave of decentralization and structural adjustment reforms as it involves promoting private businesses as the key to growth. Eventually, there will be a 'decline in the contribution of small-scale industry to total manufacturing value' as the economy grows (Nanjundan, 1989:6). As larger industries develop, the small producer may be pushed out of the market by the lower prices and (sometimes) higher quality which the mass producer may offer. For the time being, however, assistance to the poorest sectors of the population through small/ micro or medium-scale enterprise support is likely to be more effective than many of the agricultural programmes or larger-scale industrial projects promoted in the past.

Small-scale enterprise development is particularly significant for those attempting to find strategies which effectively alter the status and power of women in the poorest developing countries. For one thing, small enterprise development as a field looks at the full cycle of production and sale of a specific item. In support of any of these enterprises, many types of interventions may be considered. Projects may be concerned with estimating the feasibility of an enterprise in terms of such factors as the available inputs such as labour and raw materials or the market. They may be directed towards improving the technology of production or targeting (or broadening) the eventual market for sale. They may be concerned with training the entrepreneurs in business skills and new technologies. They may facilitate access to credit and consider profit, saving and reinvestment possibilities. They may even consider the whole chain of production and devise supports for each level involved. This multi-faceted and flexible perspective is particularly appropriate for reaching poor women. Often, these women have not been able to receive credit directly, are unable to get government-provided seeds or tools, and do not get training. They must be responsible for a wide range of reproductive tasks as well. Thus, programmes oriented to specific economic production and geared to what women are able to do with their available resources (including time) may be the only way to increase their productivity without substantially increasing their labour burden – until society undergoes a revolution and more equally shares the reproductive task load!

Limiting small-scale enterprises to handicraft production is not the intention here. The latter is a strategy which may limit women to 'women's things' and often keeps them in a labour-intensive, low-return situation. (Dhamija, 1989:195–212) The broader effort to find projects which will produce income for women based on local skills and abilities, including processing, manufacturing and other productive activities of all possible varieties, may be one of the best possible planning tools. (Grown and Sebstad, 1989:937–952; Dulansey and Austin, 1985:79–131; Downing, 1990; Carr, 1984) In the long run, major policy changes will have to occur if women are to achieve any kind of equality in the economy and politics of most developing countries. Discrimination against women in education, hiring, salaries and benefits will have to be stopped and pro-active programmes to bring women up to some kind of equality in competition in political and economic matters adopted. But, for now, in an era of dwindling resources and an international backlash of conservatism, programmes to support SMEs offer one important positive element in a somewhat ominous landscape.

This study evaluates the impacts of eight small and micro-scale enterprise projects on women in Bangladesh, India, Thailand, Ghana, Tanzania, Peru, Honduras and Guatemala. The eight countries represent a wide range of economic conditions as well as cultural expectations for women's involvement in economic production. In addition, each project had its own set of programme strategies and schedules of implementation. Some provided a full range of support such as training in group organization, business and marketing; training in the use of a new technology combined with assistance in management of the business; and marketing of the product. Others offered only one or a few supports such as venture capital for an enterprise or introducing a new technology (and training in its use). All have been funded, at least in part, by one of three donor agencies which have had numerous projects supporting SMEs: the United Nations Development Fund for Women (UNIFEM) headquartered in New York, Appropriate Technology International (ATI) in Washington, and the Intermediate Technology Development Group (ITDG) in Rugby, England. In every case, however, the three international agencies worked through local non-government organizations (NGOs), local government institutions, other international donor agencies or some combination. The types of enterprise varied widely, ranging from assorted food production for local consumption to commodity export production. The people targeted by the project had different characteristics as well. Some were already skilled in the enterprise to which assistance was given. Others had had virtually no background in any kind of business at all. The projects were also of varying duration, some having been in existence for more than thirteen years and others recently started (within the last two to three years).

Given the large differences among the projects, the task of sorting out which impacts result from the particular conditions of the project and the local political, economic and cultural environment in which it was located, and which might be attributed to the type of programme strategy, is difficult. Yet, the research is important. It provides needed data on the actual results of the now popular projects to support small enterprises. Understanding the impacts of SME projects on the lives and work of the women involved should help donors and others interested in the fate of women in developing countries realize the long-range consequences of intervening in specific ways in local situations.

In particular, this study asks whether the different types of strategies employed in each of the various SME projects resulted in different types and levels of impacts on finances, the life of the women themselves and on the family. It examines changes in the woman's income to see if it was markedly improved and whether this was reflected in the overall economic position of the family. It explores the degree to which she was empowered because of her experience and whether, as a result of the project, she now has more authority in her family and community and a better feeling about her capacities. It looks at the extent to which her life was made easier because of her participation in the project. It asks about her family and whether they received more advantages, such as better food or clothes or more education for the children. It even asks how her husband felt about her work, and whether he was threatened by her new role or pleased by the benefits accruing to the family.

Ultimately, this book questions whether projects supporting small and medium-scale enterprises have had any profound positive or negative effects or have been only peripheral in the lives of those who have participated in them. It also explores the controversial (among donors and planners) question of whether the impacts experienced by the women and their families differed depending on whether a project intervened comprehensively or simply, that is, by providing multi-faceted support such as training in technology use, business methods, self-awareness and group organization, or simply offering, instead, a single type of assistance such as access to credit or a new or improved technology. This question is hotly debated, with some agencies saying that not only are impacts about the same as those in complex projects, but also commitments of resources, and therefore costs, are much lower with the simpler projects. If true, this suggests an implied waste or even overkill in some elaborate schemes for support of SMEs although it begs the question of whether the same population of women can be reached by both types of projects including the poorest, least-trained and inexperienced women, as well as the better-off (although still poor) women who have an established SME and at least experience in running it, if not training.

A second and closely related issue which is explored here is the role of mobilization training in projects for women. What is implied here is the training of women in group organization and management, self-awareness and empowerment to assist them to take charge of their own lives and take a larger role in their economic activities and in their larger communities. UNIFEM has placed more stress on this as a project component than has either ITDG or ATI and the utility of this emphasis has been called into question by some observers. The debate is whether mobilization is either necessary or useful. Some argue that as women succeed in business, they naturally become empowered. Emphasis on group training on this dimension is therefore not needed, may violate traditions in the area and thus slow project success down, and may not even be popular with the women in the project themselves who often are not oriented by custom or tradition to women's group activities. The counter-view is that women's groups provide support otherwise lacking to individual women in very conservative environments which can be instrumental in making them successful in their work and more effective in their wider environment. Furthermore, women's groups, properly trained, can be an important resource in business management, in production, marketing or simply credit management, especially in environments where individual women do not have the resources to take out the substantial loans needed for a major new undertaking, or the alteration and adaptation of an old one.

The results from this study may have serious ramifications for those planning projects in the future. As internationally available funds dwindle, government agencies, NGOs and others query what is the most effective way to use the resources available. What will make the most difference in either the long or short run for the poorest women in the most distressed economies? This study can not claim to answer this definitively, but, it can throw needed light on the subject. SME projects have been said to assist poor women in improving their daily lives and achieving a better outlook for their near and foreseeable future. This study illustrates what these projects can and can not do and why. At the very least, evidence from this study that certain strategies do or do not produce major impacts on women's lives can be taken into account when future SME projects are devised.

Framework for analysis

The first problem which a study of this kind encounters is one of context. A single strategy – for example, offering credit – may have different impacts depending on the conditions of the women recipients. Variables in the situation include the national and regional economic context. Is the country under consideration a rapidly growing and prospering economy, for example, or conversely a stagnant and recessive environment? Is the government nationally, regionally and locally supportive of the informal sector through policies and programmes such as favourable tax laws, direct loans or subsidies to banks which loan to the small entrepreneur, and government extension programmes helping train and backstop small entrepreneurs? What is the role of women in the economy generally – are they normally expected to keep out of the public eye or are they commonly involved in economic activities outside their homes? To what degree are women usually able to control their own work or make business decisions generally? On the individual level the situation can also be extremely varied. For example, are the women who will receive credit experienced in their field of activity and are they already in an on-going business?

The first hypothesis on which this study is based is that the more favourable the external and immediate situation is for a woman's business, the less the amount or kind of intervention necessary to successfully support her enterprise. Specifically, this means that the better the economy, the more independent women generally are, the more women are normally active in business outside the home *and* the more trained, experienced and integrated into the economy a specific woman is, the less that needs to be done for her. Stated this way, this hypothesis indicates the cruelty of the development situation. Those who are in the greatest need are the hardest to reach both in terms of amount and kind of effort and, concomitantly, cost. Credit alone, then, may be quite sufficient to bring a business to a critical level of operation where there is a good long-run probability of sustainability, profitability and growth. But credit alone may not work where women have had little previous business opportunity and no support for such work either from their immediate environment or in the larger regional and national context.

Assuming this hypothesis to be true, not only credit but all possible strategies which might be involved in a SME project have to be considered from the perspective of timing: when, that is, and under what conditions they may be necessary. A further consideration is the kinds of impacts beyond the growth and sustainability of the particular business involved which will be produced by the different types of intervention, and why there is variation. A second hypothesis advanced here is that, in a successful project,[1] the impacts will be broader the more comprehensive the package of interventions. Single-factor interventions are less likely to impact on the lives of women outside their work, and less likely to have an effect on their status in the family, level of independence or overall outlook on themselves. Of course, broader impacts may be less necessary when the situation at the outset is more favourable. Thus in a country where women are already active in the marketplace, relatively independent and assured in their status, improvements in these areas need not be sought – merely strengthening the particular business may be enough. A corollary assumption of this study is, in fact, that projects in countries where women have the lowest status and the least opportunities will experience the greatest (and broadest) impacts in successful projects.

The final hypothesis considered here is that training groups of women to take charge of their own lives and work (mobilization) will have impacts which go beyond other strategies and will enhance the effects of the latter on the women in the projects studied. Thus it is assumed that women who have not received mobilization training will not have the same level of personal status, improved self awareness and outlook on their economic future or increased decision-making power either within their families or within the broader communities in which they live.

Sponsoring agencies: UNIFEM, ITDG and ATI

The eight SME projects considered here vary widely in the strategies adopted by the three donor agencies and their local collaborators. In theory, these differences should have to do with the nature of the agencies themselves, the specific goals they set for their projects generally, their perception of the most effective way to intervene, and their resources, both financial and in terms of personnel (Buvinic, 1989:1050–51). A brief description of the origin and general purpose of the three international organizations whose projects are studied here should provide some useful

---

[1] A successful project is one that helps an enterprise to become profitable and sustainable over a relatively long period.

background to a consideration of different strategy choices which these agencies have made. (The local collaborating institutions are discussed in the chapters dealing with the individual cases.)

United Nations Development Fund for Women
UNIFEM was created in 1976 by the United Nations General Assembly and is an autonomous subsidiary of the United Nations Development Program (UNDP). UNIFEM was started because of the growing awareness of the special needs of women throughout the world, and particularly the needs of low-income women in the poorest nations (UNIFEM, 1988). In its first years of operation, most of UNIFEM's funds were used for specific projects for women only. Almost half of these projects were group income-raising efforts. That is, these were projects where women of a specific village or villages or geographic zone (as opposed to individual women) were assisted in organizing themselves in groups in order to receive credit for establishing (or improving and expanding) a particular enterprise or set of enterprises, and were given training and extension services. The emphasis was on community development and assisting the poorest groups of women. The enterprises included, among others, producing and marketing handicrafts, poultry raising and textile and food production. A change in emphasis occurred in 1985, when UNIFEM began to direct more of its efforts toward long-term, indirect government policy reforms which would benefit all poor women. The charter revision of that year established the need for UNIFEM to use some of its resources to work with government agencies and ministries to this end (Buvinic, 1989:1050). By 1991, UNIFEM had assisted more than 800 projects and had 40 projects on-going in 25 countries, as well as 28 global programmes either projected or underway (Buvinic, 1989:).[7]

In 1993, UNIFEM had an income of just under twelve million dollars, of which approximately one million came from contributions from governments and half a million from women's organizations, UNIFEM national committees and NGOs (UNIFEM, 1993); the largest single contributor in 1993 was the Netherlands. UNIFEM's four major emphases included promoting women's rights, helping women gain access to credit and training for their work, studying the impacts of science and technology on women and seeking ways of making them work for women and, finally, promoting women's causes through a variety of publications. Examples of UNIFEM's promotion of women's rights activities included promoting human rights through the Beijing Conference in 1995, supporting a coalition of women at the Human Rights Conference in Vienna in June 1993 (and at the Earth Summit in Rio de Janeiro in 1992), and holding workshops or conferences for women, such as the Summit in Phnom Penh in March 1993. At these fora, women spoke out about their needs and their current situations and worked together to try to devise and propose ways of addressing current issues and problems they encounter. UNIFEM has had many specific projects and programmes together with other donors, with governments of developing nations and with NGOs helping women overcome a specific barrier in their work or in their daily lives. For example, UNIFEM sponsored a programme (together with the Danish government) offering credit and training to women in Tanzania, had a project offering training both in literacy and livestock raising in Nepal and, together with ACCION, has provided stand-by lines of credit to banks in Bolivia and Columbia to leverage secondary lines of credit for small loans to entrepreneurs. UNIFEM publications in 1993 included four books and three issues of its newsletter published in English, French and Spanish. Women, Ink. is a service funded by UNIFEM which (since 1993) has increased the availability of materials on women in development through catalogues listing what is available and where (UNIFEM, 1993).

Intermediate Technology Development Group

ITDG was the forerunner of the appropriate and intermediate technology movement which spread from England throughout the developed and developing world. The concept of 'appropriate technology' is identified with the work of E.F. Schumacher and may be defined as 'a technology tailored to fit the psychosocial and biophysical context prevailing in a particular location and period'. (Willoughby, 1990:16) This term, and the emphasis on the necessity of choosing the technology which would maximize the benefits needed by poor countries in their development process was first promoted internationally at a conference at Oxford University in 1968. Since then, the term 'AT' has taken on many meanings and has been identified with many different perspectives on technology and economic growth. However, a common core of interest in the use and adaptation of simpler and traditional technologies (which may be modified to more complex technologies as circumstances change) became the basis of institutions, usually concerned with the development of poor nations, located all over the world. Indeed, by 1980 there were estimated to be at least 1000 such institutions (Whitcombe and Carr, 1982).

ITDG itself was set up by E.F. Schumacher (and his colleagues George McRobie and Julia Porter) in 1965. The organization was called 'Intermediate Technology' Development Group because of Schumacher's emphasis on choosing the most appropriate technology through the best combination of traditional and modern technologies (Willoughby, 1990:111). The organization was oriented to helping people in poor countries help themselves, despite the clear awareness that large industrialized countries, in acting to further their own interests, had long been taking advantage of, indeed exploiting, poorer countries. A lack of resources, including trained manpower, a development research capability, and funds for investment, hampered poorer countries from using capital-intensive technologies to their own advantage. The emphasis on technology choice permitted them to review their own programmes and the plans of foreign investors to seek the best modes of production for their own development needs. As such, ITDG was in the vanguard of a movement to empower poorer nations to undertake better development planning for themselves.

Funding for ITDG programmes and projects comes primarily (40.1%) from the Overseas Development Administration (ODA, the UK equivalent of the Ministry of Development), but certain branches of ITDG, and certain programmes, also raise funds. In 1991/2, almost one-third of ITDG funds came from private donations; the rest were received from other governments, charities and groups. In 1993, ITDG was operating with a budget of £6,419,700. The professional staff has been recruited on the basis of technical expertise in one of the several fields of ITDG specialization and for their development experience (ITDG, 1992).

ITDG has always emphasized the development and dissemination of technologies. Its original method of operation was to respond to a request from an existing organization to help develop or improve a technology for some specific end. ITDG staff were mainly engineers who could provide technical services, information and input training to the requesting group of industrialists. ITDG did not have projects of its own creation. In 1983, ITDG began to change its method of operation and establish its own projects. ITDG was no longer only responding to the initiative of groups which needed technical help, it was looking for groups which would benefit from working on a technology to help them establish an enterprise or simply improve their productive processes. ITDG also supports many projects for women. In addition, an initiative was started in 1989 called 'Do It Herself' which included several conferences, workshops and media presentations all 'challenging the invisibility of women in the use, adaptation and change of tools and equipment'. (IT, 1992:8–10)

Appropriate Technology International
The founder of ITDG, Schumacher, must be given considerable credit for the interest which developed in the United States in AT and, specifically for the founding of ATI in 1976, although the new organization took a discernibly different focus from its British predecessor (Willoughby, 1990:216). ATI was created by a mandate from the United States Congress to the United States Agency for International Development (USAID) in 1977. It is a private, not-for-profit development assistance organization. Initially funded by a grant of one million dollars for one year, this amount was increased so that its next three-year government grant was 20 million dollars. ATI's 1993 annual budget was almost six million dollars, of which forty-four per cent was from USAID and fifty-four per cent from eight other donor organizations (Drake and Sullivan, 1993:34–35). Although a substantial part of ATI's funding is channeled through USAID, ATI operates independently and follows its own mandate: to experiment and

develop innovative approaches to technological development and the transfer of new technologies (Delp *et al.*, 1986).

ATI's headquarters are in Washington but its field operations are throughout Africa, Asia and Latin America. It has field offices in Dakar, Senegal; La Paz, Bolivia; Arusha, Tanzania; and Manila, the Philippines. Its projects, carried out in collaboration with selected national economic development-oriented NGOs which focus on the rural poor and small enterprise development, specifically stress the need for commercially viable and economically sustainable technologies, and so technologies developed are to be used in enterprises which would be profitable and enduring. ATI's programme focus has evolved since its earliest years from projects which emphasized the promotion of discrete appropriate technologies (through small enterprise support), to broader-focused subsector projects in which a range of interventions can be introduced throughout the subsector. ATI has had many projects which support women entrepreneurs; for example, the maize mill project in Cameroon was in part based on providing credit to women's co-operatives (and other groups) for the purchase of locally produced maize mills (De Wilde and Scheurs, 1990:25–45). Women and their needs are stressed in ATI publications, one promotional brochure underlining that 'Many of these entrepreneurs (to whom ATI is targeting its dissemination efforts) are women and their requirements and opportunities are a central focus in the field and throughout the organization' (ATI, 1993:6).

Clearly the three organizations have a multi-faceted approach to development projects in general and to SME projects in particular. In each case, the agencies have sponsored a wide range of types of programmes in which different strategies have been employed. All three have also had many programmes in which women were the targeted beneficiaries. There are some differences among the three in emphasis which may affect the project goals and the resultant choice of strategy. A primary mission of both ATI and ITDG is to disseminate appropriate and viable technologies. Women are sometimes the beneficiaries, but their promotion is not the *raison d'être* of either organization. Economic, social and legal or political promotion of women is, of course, the basic purpose of UNIFEM. It is sometimes difficult to distinguish in field projects how ATI and ITDG differ in their orientation. However, ITDG has the reputation for placing greater emphasis on developing/disseminating technologies, and ATI is known for having a greater interest in business support, with technology dissemination as one major tool in the process.

How these differences affect strategy choice is not completely self evident. *A priori*, it would seem that UNIFEM would emphasize broader project goals of overall improvements in well-being and empowerment for women, while ITDG might stress technology viability rather than a broader mission, and ATI might focus evaluation primarily on one dimension, the success of the business involved, rather than its wider impacts (see discussion in Buvinic, 1989:1050–51). In both ATI and ITDG, such a relatively narrow perspective may be the result of the belief that success in this one dimension will ultimately lead to broader life improvements for those involved. However, this possible difference in approach to project choice should result in different types of projects being chosen on average by the three organizations. The entire matter is further complicated, however, by the fact that the three organizations have collaborated with each other in many projects, and even in this study, one project (in Peru) is sponsored by both ITDG and UNIFEM.

Strategy choice in eight SME projects

In order to explore the hypotheses advanced above, this study was structured to examine a set of characteristics of the countries and the participants in the eight programmes and then to look at the different packages of interventions or strategies which were used in the different projects.The differences in strategy choice in the eight projects are presented below in Table 1.1. The full explanation of what was done in each project, why different choices were made, what specific goals were established at the outset and actually accomplished over the life of the project is recounted in Chapters Two to Nine where each case is discussed in detail. Here, only the broad distinctions relating to the classifications discussed above are presented.

Table 1.1 Strategy choice in the eight projects*

Country	Skill level	Mobilization	Technology	Training	Marketing	Credit	Initial capital
Peru	1	1	3	2	1	1	n/a
Honduras	1/2	2	3	2	3	3	3
Guatemala	3	2	3	3	3	3	n/a
Bangladesh	1	1	3	2	3	1	3
India	1	1/3	3	2	3	3	2
Thailand	1	1	1	1	3	1	n/a
Ghana	3	3	3/2	2	3	2/1	1
Tanzania	1	3	2	2	2	3/2	n/a

* 1 = low (chosen because of need) or a strategy which received no emphasis or was provided to only a few of participants. 2 = not a major stress throughout the project but considered and provided to at least some of the participants. 3 = high (or skilled), strategy a major emphasis in the project design.

Peru: ITDG/UNIFEM. Enterprise: Food Processing (various)
Honduras: Enterprise: Cashew Nut and Fruit Processing
Guatemala: ATI. Enterprise: The Wool Sub Sector
Bangladesh: ITDG. Enterprise: Dried Coconut and Sisal Products
India: UNIFEM. Enterprise: Sericulture
Thailand: ATI. Enterprise: Pickled Ginger
Ghana: ITDG. Enterprise: Shea Butter
Tanzania: UNIFEM. Enterprise: Food Processing (varied)

The table is organized to show what strategy choices were made in the projects. 'Skill' refers to whether participants were chosen on the basis of existing experience in a particular enterprise or because of their poverty and need. Whether training was broadly conceived to include self-awareness, empowerment and group organization beyond the tasks required by the specific project enterprise is the second strategy and is included in the table under the rubric 'mobilization'. Introduction of a new or adapted technology is the third classification. Training in technology use with some discussion of enterprise management is a fourth category. Assistance in marketing is a fifth. Provision of credit to the participants is the sixth choice, and the seventh is provision of initial capital without requiring the client to understand that she is repaying a loan.

The table shows some interesting comparisons, although it has certain inherent problems. For example, the project in Thailand was a venture capital project where money was made available for investment in an enterprise (a pickled ginger factory). In this case, capital was provided and more leveraged to add to that which an initial group of entrepreneurs already had available. The beneficiaries of this project, therefore, could be these entrepreneurs, the traders who purchased ginger, the poor farmers who sold ginger to the traders or the factory or the poor men and women who provided the labour in the factory. In this table, the poor farmers/labourers are considered the beneficiaries but this does not do justice to the intent of the project. Equally confusing are the double numbers meant to indicate that the emphasis was either not followed throughout or not given uniform stress. Thus, changes in mid-project in India occurred as a result of UNIFEM's perception that mobilization of the women was necessary, but only a few women had been reached by this strategy shift when this study was done.

A more substantial problem results from the necessity to select a few important strategy alternatives to consider as points for comparison as opposed to all possible components. Furthermore, the chosen strategies themselves could not be systematically broken down and contrasted according to all their relevant possible variations within the table. For example, training in use of technology is highlighted here, but training in business skills (and backup in this area) are not separated out. Clearly, business training is in itself an important project component which can have its own discernible impact. Thus, too, as an example of possible variations in any strategy category, 'mobilization' could take many forms. It could be long-term intensive, interactive sessions with leader-guided group role-playing so that women could get experience in taking charge of problems which arise in their enterprises and in their daily lives generally. On the other hand, it could be a few sessions of discussion about the importance of women co-operating together in a group with some technical explanations of a group undertaking such as the management of a loan or the use of a particular technology. Mobilization will also have different content and different implications if women have already had such training earlier.

Despite the value of more detailed information on strategy components in the eight projects (which is included in the individual case discussions in Chapters Two to Nine) this table, for reasons of space and clarity, captures only a portion of the information on how the eight projects were organized and stresses only seven strategies as a basis for comparison. However, Table 1.1 does show some important points. First, no agency sponsored only one type of project. ITDG certainly maintains a primary emphasis on technology/technology transmission in these projects. In addition, if Peru is considered a ITDG project primarily (ITDG planned and organized this project), UNIFEM seems to stress mobilization more than the other two agencies. ATI has only two projects under consideration but in both the business/profitability emphasis is clear. Overall, the diversity of the eight projects is so large as to prevent further attributing of the strategy choice to initial agency orientation. It is possible, of course, that agency orientation will affect the manner in which the projects are run and the criteria used for evaluation, but these matters can not be raised until the individual projects – and the local partner organizations – are considered at length.

The table shows that all but two of these eight projects should be classified as moderately or extremely complex, that is, having a multi-faceted, relatively comprehensive set of supports. Except in Peru and Thailand where the projects stress the provision of one input (technology and capital respectively), the agency commitment is moderate to high due to the complexity of the projects and the various skills needed by agency staff. In all but two cases (Ghana and Guatemala), the project participants were chosen on the basis of their need rather than their pre-existing business experience, so training is a necessity. In other research, observers have found a direct relationship between degree of complexity of a project, associated with degree of agency commitment (and cost), and the extent to which the poorest, least skilled people can be reached. They also suggest that risk of failure increases with increasing complexity (Ashe, 1985:27–36; Mann, Grindle and Shipton, 1989:53–67). In our research we want to go further to explore the impacts of these different strategy choices on the women who participated in the eight projects. Are impacts equivalent, despite different packages of inputs, if projects succeeded in creating financially viable enterprises? How viable in the short or long term have these projects been, and does this relate to the strategies adopted or does it seem to have been an artifact of the specific national situation?

Ranking the countries

The hypotheses for this study assume that significant differences in national and regional economic circumstances and the degree of access of women to full participation in the economy are important in determining whether

Table 1.2 Ranking of cases by contextual factors (highest = 1)

Level of Development	Women's Participation in the Economy
1) Thailand	1) Ghana
2) Guatemala	2) Thailand
3) Peru	3) Tanzania
4) Honduras	4) Peru
5) Ghana	5) Honduras
6) India	6) Guatemala
7) Bangladesh	7) India
8) Tanzania	8) Bangladesh

specific types of intervention are needed and what impact they will have if they are used. Table 1.2 ranks the eight cases for both these factors.

Table 1.2 illustrates the problem of comparing women's projects across countries and regions because the question of what constitutes women's access to the economy takes on a slightly different connotation in different cultural contexts. The level of economic development, in contrast, is relatively easy to measure and can be taken from a number of international statistical compilations. In this table the 1994 World Bank ranking of countries by basic indicators such as GNP, rate of inflation, life expectancy and illiteracy is used (World Bank, 1994:162–3). Common sense might suggest that the level of economic development should be directly correlated with the degree of access women have to the economy; yet this is not the case. What controls women's access to the economy is the amount of discrimination against women in any given society, determined by such factors as national and local traditions and religion, current waves of conservatism or liberalism *and* the rate of the growth of the economy. A single composite measure made up of all of these taken together is hard to develop, which makes comparison very difficult. A factor related to this composite measure is certainly the extent to which women have access to formal

Table 1.3 Access to education and participation in the labour force

Education of Women (% Females per 100 males in Primary School 1991)		Female Share of Labour Force (% in 1992)	
1) Tanzania	98%	1) Tanzania	47%
2) Honduras	98%	2) Thailand	44%
3) Thailand	95%	3) Ghana	40%
4) Peru	*	4) India	25%
5) Honduras	*	5) Peru	24%
6) Ghana	82%	6) Honduras	20%
7) Bangladesh	81%	7) Guatemala	17%
8) India	71%	8) Bangladesh	8%

* No information given.
Source: World Bank, 1994:218–19.

education and another is the extent to which women are integrated into the wage labour force. If these were the criteria selected for this study, however, the ranking in Table 1.2 would be different. Table 1.3 shows the eight countries studied here by these factors. The two lists are quite different from the two above and also not consistent with each other.

To derive the list presented in column two of Table 1.2, the region of each country in which one of our cases was located was looked at in terms of how women were viewed by their society – were they permitted to seek work outside their homes on their own initiative? Were they relatively equal to men in their freedom to decide on the use of the income from whatever work they did? Were they considered out of place in the public sector or, that is, were they as free as men to go where they wanted? Who directed household affairs? These questions are harder to answer without interview and survey data available. In this study, one aspect of the interviews and background information which was collected concerned this question directly. The list presented in Table 1.2 reflects the results of this analysis.

Ultimately we find strong economic-level differences between the first and last countries (Thailand and Tanzania) in levels of development, and equally strong differences between the countries where women have the greatest access (headed by Ghana) and the countries where they have the least (Bangladesh being the lowest on this scale). There is a regional factor in that the south-east Asia cases (but not Thailand) present the most difficult conditions for women to participate in work; the Latin American cases group together in the middle of both scales, while the African cases indicate better access for women but are not consistent (or strong) in the overall measure of economic development. Thailand alone comes out at the top of both lists, having the strongest economy and the best conditions for women's access. These differences strongly affect the way different strategies chosen in various projects may affect women. They will also be important in determining how – or if – certain strategies may be successfully applied, as the following chapters illustrate.

Access to education for women and the extent to which they are integrated into the wage labour force are also important in evaluating the overall situation for poor women but, since they are determined in large part by the underlying attitudes and customs which we have tried to capture here in column two in Table 1.2, they are not considered as determinants in our study. However, we do look at whether and to what degree the women participants themselves are educated or trained. Their level of education should influence the way they understand and participate in a project, and also the type of strategies which may be appropriate, and what impact such strategies may have. We also consider their religion and whether or not they are heads of their household (and/or single) and how old they are. No one religion can be classed as more or less supportive of women's access to economic participation; it depends on the context in which that religion is placed (Callaway and Creevey, 1994). In a given group of women in the same setting, however, type of religious belief may also affect why one set of women responds in one way and another does not. Being single may mean a greater degree of independence but it also may mean fewer resources. Thus this characteristic is important to consider. Age, too, affects what women can and can not do. Older women are not necessarily less flexible or adaptable to change. They may be, however, just as they may be freer than younger women still bearing and raising children. Thus this too needs to be taken into account.

This book describes the set of national and individual characteristics for each case and the project characteristics and experience in each country. Then, in the final section, the experiences and results from each project are placed within the context of a matrix of these variables and the strategies used and impacts sought, and analysed in a set of quantitative cross-national tests.

Research methodology

To explore the basic assumptions in this study that the more developed economic environments required less complex interventions for small enterprises to succeed, and that women with established businesses and some experience in the field also required less complex interventions than those with no previous training and without established SMEs, a careful research design was needed. This was made the more difficult because the assumption that the kinds of impacts would vary depending on the type and complexity of intervention had also to be investigated. Ultimately, the study was based on the results of two basic types of questionnaire, one for individuals who had participated in the study or a control group of individuals with similar characteristics who had not been included in the project, and one to be given to owners/managers of enterprises to explore the specific impacts on the businesses assisted by the projects. The basic questionnaire was slightly altered to suit the different circumstances and administered to smaller groups of family members of men or women who had been in the projects.

The questionnaires

The first questionnaire addressed individual issues of four major types:

1) The socio-demographic characteristics of the respondent including age, sex, marital status, number of children and their education, rural or urban dwelling, education and other training, participation in social and economic groups, similar characteristics of husband (or wife), etc.
2) The economic status of the respondent including income earned, assets held, economic activities (with considerable detail as to what tasks they

performed, not only in the enterprise but in other economic activities) and the activities, income and assets of their husband (or wife) and how their income (and that of their spouse) was principally spent. In addition, changes over the last few years in these dimensions were also queried.
3) The decision-making patterns in the respondent's family: who decides whether the wife works, what she does, how her income is spent, where the family lives, how either the boy or girl children are educated, etc., and whether this has changed in the last few years.
4) The participants' evaluation of the project's impact: what it had done for them (credit, training of all kinds, extension and management support, marketing, etc.); how it had affected their income, assets and expenditures; what kind of impact it had had on their authority in the family and their self-esteem; how it had changed their daily lives, specifically in terms of their use of time for the household, childcare, their own leisure, education, and economic activities; what their husbands thought of the project; and how they felt it had affected their family, specifically in terms of the family diet and children's education.

The second questionnaire delved specifically into the organization and operation of the enterprises, including the number of employees, their gender, training, salary and responsibilities, the range of products, the method of production, the changes in economic conditions and the profit/loss ratio over the years of the project. The owner/managers interviewed were also asked to assess the impact of the project on the enterprise.

Research was carried out by a team of eleven women. The principal investigator (the author) developed the research plan and remained in the US to co-ordinate the team, handle difficulties which arose, pursue library and documentary research and, finally, to take all reports and questionnaires, process and analyse these and write the final conclusions of the study. Two of the team, Carolyn McCommon and Valerie Autissier, were the field supervisors. They went to the eight countries to acquaint local scholars with the goals and research plan of the study. They also introduced them to the questionnaires and trained them in administration of these to maximize conformity in the way responses were gathered and interpreted.

Before the two field supervisors went to the eight countries in 1993, a tentative designation of the samples and the method of their collection was determined. The procedure adopted was a multi-stage random sampling process. In each project, the major differences among participants and the overall range of enterprise experience were used as points of stratification of the sample. Criteria included geographic location (i.e., representation of rural–urban, regional or village-to-village differences in the project, etc.), variety in the types of enterprises, length of experience with the project, and type of involvement with the project (i.e., in some projects, there were different types of participants including home workers as opposed to factory workers (Bangladesh); or farmers (sellers), traders and factory workers (Thailand); or sheep farmers as opposed to weavers (Guatemala), etc.) The eight local researchers – Marisela Benavides, Ana Ruth Zuniga Izzaquirre, Luz Marina Delgado, Afsana Wahab, Ila Varma, Rachittas Na Pataling, Mary Abena Kyei and I.H. Mafwenga – met with the field supervisors when they arrived in their countries. Together with the in-country project staff, one field supervisor and the local researcher reviewed the tentative stratification proposed for each case and revised it in terms of the existing actual project conditions and on-going changes which affected participants of the project and distinguished among them. Each local researcher was asked to administer specific numbers of questionnaires to chosen sub-sets of participants (determined by proportional representation in the project in most cases, although some over sampling was necessary to gain a perception of a needed perspective), starting from a fixed point and randomly interviewing at every third household (or fifth in an urban setting) until the required number of questionnaires was reached. A minimum of thirty questionnaires overall was requested for both the sample of participants and the control group.[2] The specific enterprises studied through the second questionnaire were chosen as representative of the variety of enterprise experience within the project.

The final sample of the individual questionnaires drawn in all eight countries differed slightly from the planned distribution owing to field conditions, last minute information on project changes, overall accessibility of participants and available time for the local researchers. Deviations which reduced the number of participants or control sample also reduced the

Table 1.4 Questionnaire distribution

Country	Female participants sample	Male participants sample	Female control sample	Male control sample	Female family member	Male family member	Enterprises*
Peru	35	0	30	0	5	8	13
Honduras	36	11	30	0	4	10	8
Guatemala	3	30	0	30	22	0	12
Bangladesh	34	0	32	0	0	9	9
India	44	0	28	0	0	10	15
Thailand	36	10	19	0	0	5	1
Ghana	40	0	40	0	0	14	8
Tanzania	49	0	40	0	0	15	8
Total	277	51	219	30	31	71	74

* Number of enterprises to whose manager/owner the second questionnaire was administered.

---

[2] Thirty questionnaires are considered a minimum sample to ascertain 'normality,' hence validating statistical tests of the various hypotheses explored in the text.

validity of the statistical comparisons. However, great care was taken to correct for small size bias with other sorts of information collected on each project. The other smaller samples (of family members, etc.) were used only descriptively as suggestions of types of opinions, reactions, etc.

The sample of completed questionnaires included a total of 679, divided as shown in Table 1.4 above.

The problem remains, of course, that the samples of women inside and outside the projects across the eight countries are not comparable to each other. The projects themselves vary widely. The type of participant in terms of culture, outlook and criteria for selection is dissimilar, and the control sample, matched in each case appropriately to the characteristics of the project women, must consequently not be comparable either. Thus, this evaluation of impacts of different strategies can only assess with certainty the degree of change observed in each dependent variable (e.g. in income, assets held, time use, outlook on self and own economic future, and family access to food and education) within each project with its own specific bundle of strategies. It is not possible to distinguish definitively among the relative impacts of different strategies where multiple ones are combined within one case. But across the whole set of projects where the combinations differ, overall conclusions can be drawn – albeit gingerly – about the relative effects of different strategies. The points of rigour in the overall comparison include the method of selecting the participant and control samples, the use of identical questionnaires to ascertain impacts (translated to make them equivalent in different cultural terms), and the presence in each country of one of the field supervisors to train local researchers.

Because there are no consistent baseline data, it is impossible to directly assess changes over time in the status of our dependent variables, the impact measures. As an alternative in both the experimental and the control groups, the interviewees were asked to remember the status of their firms, and their living situation, and the changes which had occurred in each over time. Because of the notorious unreliability of memory as a basis for fact, this method is far from ideal. Nevertheless it provides some basis for comparing the perceived rate and level of change. The quality of our control group is, once again, critical. We can, with some degree of confidence, assume that the memories and biases of the participants and the control group are roughly comparable. Thus, differences between the two in terms of recollected change may be attributed to the project impacts.

Supporting research

In addition to the information provided by the surveys, substantial background data was required to explain the conditions faced by the participants and their project experiences. The local researchers were asked to provide the following types of material:

1) The inflation rate for the country and region over the time of the project up to the present, and the amount of devaluation of currency if this occurred.
2) In regard to women in the rural project areas: what kind of family authority would a woman typically have (distinguish among the various ethnic groups if this is relevant)? Could she maintain a separate account for her economic activities? Would she normally decide on what she does in her work and what kinds of assets she acquires? Can she decide on the use of her salary or other income? Can she own (have registered in her name) property and is that common or uncommon in this area? Is there a tendency for male family heads to become more involved in the women's economic decisions as they become more successful?
3) Government policy toward small and medium-scale enterprises. Is there a policy, that is, are there any government programmes or actions to support (or discourage!) small/medium-scale enterprises such as making credit available (either directly or by supporting banks), grants, extension agents who help small and medium-scale entrepreneurs, pricing, tax or other financial policies which support such enterprises or any other relevant supports?
4) Events which might have affected the project such as droughts, floods, extreme cold, extreme heat.
5) The existence of blights, plant or animal diseases which might have affected the project during the project period.
6) Relevant political or social factors such as as civil disturbance or unrest, opposition of a local or regional leader to the project, opposition by one or other ethnic group to the project.
7) The average per capita and family income for the country and the region of the project.
8) Questionnaires pick up certain kinds of information but the informal reactions of the participants and others in the region and even in the national capital to the project may be revealing. The researchers were asked: what reactions do you hear? Is the project well-known in its area? Is it liked or envied (or the opposite) and why?

All eight local researchers maintained notebooks and commented extensively on the responses of the participants and those in the control sample. They interviewed project staff members and, on occasion, conducted other in-depth interviews with participants which went beyond the questionnaires. Carolyn McCommon and Valerie Autissier provided extensive reports on all projects they visited, together with notes and documentation, information from interviews with project field managers and from staff in the three international agencies, and observations from their own field interviews with project participants. All interviews, the administration of the questionnaires and supporting research were completed in September 1993. The data was coded and processed in Storrs at the University of

Connecticut. The final draft of the report, upon which this book is based, was completed in June 1994. In October 1994, UNIFEM held a workshop at which the findings of this study were discussed by a panel of experts with long experience of projects supporting women's enterprises. Results from this session were used to shape the final form of this presentation.

Guide to the discussion

This book is organized so that each case may be viewed in detail, starting with some background information on the country, its political, social and economic conditions, the situation for women in that country and the general government policy towards SMEs. The projects are briefly presented in terms of their original goals, strategies adopted, funds received and organization of implementation (from the international donor agency down to the ground-level project) and actual experience. The question of project impact, as indicated by the survey results, is then considered. Here, the general developmental goals for women which underlie this study determine which impacts are examined. These include: improvements in individual and family well-being and in individual and family income; increases in the women's responsibility, status and sense of self worth; and augmentation in their authority in the family and community. All of these goals may – or may not – have been part of the original project objectives but this study has adopted its own broader framework for the purposes of this research.

In Chapter Ten, an examination of the entire cross-national sample is made to see what patterns appear most strongly to distinguish projects (and their different bundles of strategy components) and their perceived impacts from each other. Here, the frailty of statistical analysis on data of this type and quantity is the major risk in interpretation, but some statistically significant patterns and trends do emerge which have intuitive sense and are worth presenting. Although we do not claim any definitive results which would eliminate the need for further research and reflection, the findings suggest key issues which merit further consideration and possible action.

Setting the bias out front

The problem of inherent biases in studies such as this, stemming from value-laden assumptions used in research and interpretation, must be raised. Since these can not be eliminated, some of the major assumptions included in our broader developmental goals for women are presented here as a clarification – or a warning – of what kinds of values were incorporated when this research was designed. It is significant that many of these values are called into question or modified by findings and interpretations based on the comments of the participant women in the analysis presented in this report. (Many other less controversial hypotheses provide the underpinning for the numerous other matters queried by this research, but these will be presented as the individual cases are presented.) The more debatable assumptions are stated here:

1) Well-being for women participants includes having *more* time to sleep, rest, and simply to have leisure time available, or to educate themselves. Projects should, therefore, aim to enhance this aspect of women's lives. (The assumption is that women work too hard in their dual domestic and economic roles and are over burdened.)
2) Well-being for women participants includes spending less time on domestic tasks (caring for the household, gathering fuel and water). (It is recognized that some women might prefer having more time to be at home.)
3) Well-being for women *may* include spending less time on child care than previously for the above reasons. (This may also be a negative if the children's well-being, i.e. health, sense of self, etc., lessens as a result.)
4) Improved status for women includes increasing their authority in the family. It may be measured by the extent to which they, as opposed to their husbands (or others in their family), make decisions about their own work, incomes and family matters such as where they live and what kind of education their children receive. (One question here is, does this then mean that a single, divorced or widowed woman has greater status than a married woman? Most participants would disagree, although such women usually have a greater degree of independence than do married women.)
5) Improved status for women includes increasing their ability to contribute to family needs. (But this may result in greater interference by their husbands in their business and in the decisions on what they do with their incomes.)
6) Project success must include some measure of long-term enterprise sustainability. (But short-term projects might result in the acquisition of skills and self confidence which would benefit the woman in the long run.)
7) Mobilization of women in groups to undertake responsibility for development schemes, or to run enterprises, with a concomitant emphasis on the empowerment of women to give them a clearer perspective on their own self-worth, their own rights and their own abilities is a desirable objective. (This may be a strategy in itself or the outcome of other strategies, which is one possible confusion. An additional caveat is that some observers feel group activities can be emphasized too much, especially where economic activities were traditionally individual or family based.)
8) Women will not benefit from projects directed only at their menfolk as much as they will from those in which they have a key role. In most

developing countries, the family system needs to be viewed as an interactive unit in which its various members jockey for position. Men and women in a family frequently keep separate track of their own work and their income. Gains for one member may mean loss in power or authority or even well-being for another despite the fact that the family may present itself as a unified whole to the world outside. (Yet women certainly may benefit in some ways from successful projects in which their families are involved. The question remains, however, how do they benefit and how does this contrast with women who are a project's target participants?)

Conclusion

This book takes a series of small enterprise projects and analyses them. The strategies adopted in the eight cases studied here were chosen by the project organizers to minimize risk and maximize success for the women participants. But the results – and the levels of success – were not all the same. What makes them work or not work is highlighted in the individual project analysis. Finally, this book considers what strategies have proved most effective *overall* in what combination and in what circumstances, and what the implications of these findings are for those working in this field. Ultimately the lives of poor women and their families must be improved or there is no development; then intervention is nothing but a costly hoax leading on the dispossessed and leaving them more bereft than they were before. To help bring about this development, then, must be the goal of this study.

CHAPTER TWO
Food Processing in Peru

Setting the context

Peru is a nation of 21.7 million people in a land area of 1 285 000 km². Average per capita annual income is US$950 but this figure is much lower outside Lima (World Bank, 1994:162). In two sites of this project, Huancayo and Huacho, the average *annual* income is estimated to be only $174 to $298. In the third project area, Pucallpa, even this figure is too high as the population lives within a primarily subsistence economy. Peru's economy has been struggling for many years. The economy had been based on commodity exports – cotton, sugar, wool, coffee and oil, and latterly, fish products, lead, zinc and tin – but these have lost their place in the world market (except in times of war or other international crisis). Peru's current principal source of revenue is manufacturing and industry, followed by the service sector. Agriculture provides only 30 per cent of the GDP (World Bank, 1992:222). Peru's annual inflation rate since 1965 is 233.7 per cent (World Bank, 1992:258). The currency was devalued in 1989. Average inflation over the period 1988–92 was 2406 per cent and Peru's external debt is now 21 billion dollars (McClintock, 1989:372) The economic decline has been so severe that by 1985, real per capita incomes were at the same level as in 1965 (World Bank, 1992:278).

Suffering from a lack of arable land, the population has migrated from the impoverished rural areas, and Peru now has 70 per cent of its population living in an urban area. Lima alone has more than 29 per cent (Stearns and Otero, 1990:140, 149). The lack of growth of the formal economy has caused a huge growth in the informal sector. It is estimated to account for 33 per cent of the economically active population in the nation, while the overall (rural and urban) informal sector has 47 per cent of the economically active (Stearns and Otero, 1990:141). In this context, then, the decision of Alan Garcia, the leader of American Popular Revolutionary Alliance (APRA) who came to power in 1985, to support the informal sector is very significant (Stearns and Otero, 1990:138–152). The government established a new NGO, the Institute for the Development of the Informal Sector (IDESI), through which government and other funds were to be channeled. The Sector Guarantee fund (FOGASI) was established to support loans to microenterprises. Garcia set up programmes for credit provision through the national bank and other government institutions (Stearns and Otero, 1990:149). He also sought to keep down inflation, and reduce the cost of basic services such as electricity, water and gasoline. He attempted to increase real salaries in the formal sector to increase buying power for informal sector goods.

Initially, Garcia's programme had an impact, including an increase in employment and income in the urban informal sector (McClintock, 1989:372), but his economic recovery plans could not be sustained as they were based on spending Peru's foreign exchange reserves and making few international debt repayments.[1] The government's ability to support the informal sector declined as the economic crisis worsened. The new president, A. Fujimori, who came into office in August 1990, maintained and extended the positive attitude and policy towards the informal sector which Garcia's government had begun. Fujimori's government implemented stabilization and structural reforms which have recently begun to show significant results. The general fiscal imbalance was reduced with substantial reforms in foreign trade and the agricultural sector. Legal and institutional reforms were initiated to stimulate domestic and foreign investment by liberalizing and de-regulating the economy. By the end of 1991, the country had finally realized a slowdown of inflation to less than 139 per cent per year, as opposed to 7458 per cent per year in 1990 (McClintock, 1989:372). But the economic situation was still extremely difficult. The government's attempts at restructuring inflicted severe hardships on the population. Subsidies were abolished in 1990, thereby introducing price increases of more than 3000 per cent for petrol and of between 300 and 600 per cent for basic foods. Few resources were available to support the favourable legislation which had been introduced to facilitate the development of small enterprises and other informal sector activities.

The principle of government interest in and support of the informal sector had been established and can not be discounted. In other countries, lack of government sensitivity to the informal sector is a major deterrent to the success of small enterprises. This is not the case in Peru, at least not since the mid-1980s. Favourable attitudes were consolidated in Peru by the publication of Hernan de Soto's influential book, *The Other Path*, which underlined the importance of the sector and indicated that a large percentage of the population of the country was dependent on informal sector work. As a result, certain barriers to the formation of microenterprises have been eliminated and certain formalities simplified, while increasingly credit is being made available – at least in principle.

---

[1] The US government withdrew disbursements of all non-humanitarian aid in April 1992 following President Fujimori's suspension of constitutional government.

The favourable governmental attitude to small enterprises is offset by the continuing economic crisis, and the closely related political instability reflected in the power of the Maoist organization, Shining Path, and the second guerrilla organization, the Tupac Revolutionary Movement (MRTA). Certainly the economic situation is at least in part to blame for hostility to the government found in many parts of the country. There is widespread sentiment that government policy has benefited only the élite and a small new middle class without helping the poor significantly. In particular, this is felt outside the Lima metropolitan area and, more especially, in the highlands to the south (where part of this project is located). As many as 5000 Peruvians acknowledge membership in the Shining Path, but Peruvian observers report that the movement is active in 70 per cent of the country (Vargas, 1991:54). Between 1980 and 1987, over 10,000 people died as a result of guerrilla violence and government reprisals (McClintock, 1989:366). One report indicates the cost of the violence of the insurgency campaign over the ten-year period ending 1992 was approximately fifteen billion dollars (Europa, 1992:2199–209).

Women in the Peruvian economy

Women have a second-class role in the economy and society of Peru, although their status and economic activities vary in the extremely different cultural zones of the country. Nationwide, women receive less formal education than men; 10 per cent of women aged 15 to 24 are illiterate, while only 3 per cent of the men are. For those older than 25, 35 per cent of the women are illiterate and 13 per cent of the men. Most children go to primary school but girls are less likely to go than boys. Almost two-thirds of all children go to secondary school, but again, girls are less likely to go than boys. Females are less likely to go to college than their male counterparts, of whom less than one-third attend in any case (UNIFEM, 1991:52; World Bank, 1992:274). The position of women in Peru is perhaps better illustrated by their lack of access to the formal sector. Only 25 per cent of women over 15 are said to be economically active, although this figure disguises their activities because subsistence farming and very small-scale entrepreneurial activities are often left out of government statistics. In contrast, 79 per cent of men of the same age are economically active. In Peru, in the formal sector, women are only 9 per cent of the administrative and managerial class of workers and 13 per cent of the largest group, the labourers or unskilled workers. The only class of employment where women have a significant place is in clerical and service positions where they are 60 per cent of those employed (UNIFEM, 1991:106). Women constitute 34.5 per cent of those in the Peruvian informal sector; 70 per cent are in commerce, 22 per cent in production and 8.2 per cent in services (Reichmann, 1989:134). However, as in most Latin

American economies, women in the informal sector tend to be in the smallest, least skilled enterprises with the lowest economic returns.

In the three rural areas where the projects under study here are located, women have less power in the household than do their husbands (or fathers if they have not yet married). General studies on Peruvian women have shown that even in the allocation of limited family resources for their children's futures, women do not have the decisive role, men do. They can not control significant rural problems such as male drunkenness, which leads to wife beating and the loss of family resources. Furthermore, while women are often involved in family farming and animal husbandry, they do not get the full benefit of their labour (Bourque and Warren, 1981:114, 105–13). Both men and women in Peru recognize that rural women are involved in agriculture and animal husbandry, but they give different weight to that fact. Women see themselves as centrally involved while men see them as peripheral aides in the production process. Men's work is valued more highly than women's and both sexes explain this as the result of men's strength and ability to use heavier tools. Women have more opportunities in trading towns, where they may get jobs in shops or other service positions, or engage in small production enterprises. However, they are paid less than men and men still dominate certain key aspects of trade and commerce even in the informal sector (Bourque and Warren, 1981:118–31; Arias and Ickas, 1985:266–70).

There is a women's movement in Peru seeking to improve the position of women, which by 1989 had already made itself felt by both government and society. This movement has been fed by the increasing access to education of Peruvian women and by the pattern of increasing urbanization in Peru, which means more women are in closer proximity to women with different backgrounds and experiences. Political socialization is thus easier and more likely to occur. Women's groups have begun to proliferate in towns and in the rural areas. Some of these are sponsored by church-related organizations. Some are responses to incentives of credit and support possibilities made available by government agencies and NGOs. Thus, there are such organizations as mothers' groups, which may sponsor soup kitchens or childcare programmes or commercial activities by the members. There are also production and marketing co-operatives for small enterprises, often supported by a specific project or programme. But there is no one women's movement; rather many different groups, some of which do not co-operate with the others. One 'stream' is educated middle-class women with a strong left-wing orientation, typified by the Social Front of Women and ALIMUPER (Accion para la Liberaćion de la Mujer Peruana), and emphasizing government policy reform and major social change. Another is the 'popular stream', emphasizing specific projects to help poor women in the subsistence sector without stressing social reform. A third stream of the women's movement may be seen to be the strong women's contingent in the Shining Path where the leadership has from the outset included women, although in a traditionally subordinate position (traditional political parties in Peru had not included women in positions of responsibility). None of these groups, however, motivates a national response among Peruvian women. And, as mentioned, some groups do not co-operate with others. Leaders of the educated middle-class stream report attacks and even killings of their members by guerrilla groups, sometimes headed by women (Vargas, 1991:55, 15–55).

In the meantime, women in the different zones of Peru cope with economic and political crisis and a slowly changing social structure as they continue their daily work. In each of the areas, their patterns are different. Looking only at the three project areas, we find these different scenarios: Huacho, a small town with a population of about 40 000, is on the coast and closest to Lima, two hours by road. It is located in the arid Huaura Valley, which stretches along the coast and is surrounded by a coastal desert. This is the most urbanized region of Peru and the one in which the greatest changes have occurred in the rural as well as the urban sectors. Land, originally held by large owners, was divided among the former workers of the large farms at an average of four hectares a family in the land reforms of 1968–74. Water is available so planting is possible all year round. Corn, cotton and sugar cane are common crops. Despite the availability of water, farming is not intensive and productivity is not high, due in part to the lack of available credit and the instability of agricultural prices. The zone is no longer primarily engaged in subsistence agriculture (in which women usually participate almost equally to men) but is divided into large commercial and specialized farms relying on paid labourers, many of whom are migrants. Most of the labourers are male, and the women of the zone generally remain at home. If the family has rain-fed crops, these are usually cared for by the men. The women maintain household gardens, help with the care of family animals (such as sheep, goats or poultry), and process the food for family consumption. The local capital, Huacho, has been severely affected by the economic crisis. Its buildings are in decay and its population is poor. Both men and women migrate to Lima looking for work although the current economic situation means that few can hope to find jobs in the formal sector.

Huancayo is the capital of the department of Junin located in the central Andes. It is one of the larger towns in Peru with a population of over 200 000. Although further from Lima than Huacho and at an elevation of 3400 metres, it is the regional centre and closely linked to the capital region by road. This is a principal source of food for Lima. The elevation determines cultivation; the lower areas have richer soil and produce primarily potatoes and grains. Families living at higher points grow various grain crops. Above 3700 metres, less and less may be grown, and the common crop is potatoes. The project villages are close to the town of Huancayo at the lower elevation.

In Huancayo, people are organized into communities but possession and use of land is individual. Each family has approximately two hectares.

Poultry and small ruminants are also raised and people conduct small artisan and commercial activities. Work on the farm is done by the whole family, women as well as men. However, when the men migrate to Lima or another town, as is often the case, the women are in full charge of the family farm. Tasks are clearly divided by gender in families where the male is still present. For example, women sow the crops, prepare them for storage, store them (for home consumption) and do the family marketing (and also selling of family products). Women have distinct family responsibilities and authority, but men are the family heads. On most farms, some family members try to supplement the farm production with outside paid labour (Rios Varillas and la Cruz, 1981:10). Because of the close linkage with Lima, acculturation is relatively widespread, and many women speak Spanish (as opposed to just the local language).

The third area, Pucallpa, is located in the tropical Amazonian region of Peru, east of the Andes and south of Lima. It is by far the most isolated and traditional of the three zones. Its character is defined by the long Ucayali river, along which 65 native communities are situated. Pucallpa, the regional capital and main river port has a population of 120 000. It is a commercial centre and also the centre for the production of wood. Most of the surrounding rural population are Indians living on subsistence farming. The project site is located in the poorest community, the Shipibo-Conibo. They live by subsistence agriculture and maintain native Indian traditions. Income derives from traditional crafts produced by women, including pottery, cloth and embroidery. This is the poorest of the three project zones; levels of nutrition are low and life expectancy short. Women have an average of four to five children. Interestingly, women in this zone have relatively high prestige and autonomy in society, especially the Shipibo Indian women who are the beneficiaries of the project. They control their own small production units, sell what they produce and decide on the use of their income (Stocks and Stocks, 1984:73–5). The women, however, are very traditional in their orientation, unable to speak Spanish, unacquainted with ideas of reform and change for women, and wary of strangers.

The Peru Food Processing Project

The food processing project was developed in 1987 to meet a crisis situation for some of the poorest families in the country. In Peru, where national resources have been concentrated on the Lima metropolitan area, rural farmers have become increasingly impoverished. Fifty-four per cent of the rural inhabitants are in fact malnourished. ITDG staff saw this situation in large part owing 'to low productivity due to a lack of resources and available technology' (Rios Varillas and la Cruz, 1981:1–5). The solution suggested by the government was to raise productivity, but this was not an adequate strategy. Farm production throughout most of the country faced a very inelastic structure of demand; when crops were plentiful, prices plunged. Thus ITDG proposed a different strategy of processing foods for availability at non-peak times of year (for home consumption as well as sale), and allowing for transport and sale outside the region without spoilage. The latter goals were researched by various NGOs and international agricultural research centres. Specifically, the production of various flours was explored. However, efforts were hampered by the costs of production, incompatibility with local food habits and technical inefficiency.

ITDG, in contrast, proposed developing technologies based on traditional production methods which could be used by the rural farmers in their own areas. These technologies would be appropriate in terms of the resource and knowledge base of the farmers and their own preferences, would result in greater profit than traditional technologies, would be versatile and able to respond to changes in products and prices, would not demand high-quality fresh products, and would be affordable by families or groups of people (Rios Varillas and la Cruz, 1981:1–5). The project would establish three demonstration workshops in three distinct agro-ecological areas so that each centre could investigate and develop a wide range of representative technologies based on local materials. These were zones where ITDG had had food processing projects and contacts with NGOs working in rural development. These workshops would have a threefold purpose:

1) to study different food processing technologies and develop appropriate equipment for the local conditions;
2) to provide a processing service for the local community which would also act as a demonstration so that local people could learn about the use of the improved technologies;
3) to provide training in the use of the technologies and help in establishment of processing plants by villages or other communities or groups (Rios Varillas and la Cruz, 1981:6).

ITDG proposed to equip the three workshops, suggest the food technologies to be further explored, and train and advise the local partner NGOs in the use of these technologies. ITDG would also finance a food technologist for each centre. ITDG/Peru (one of the oldest of the ITDG offices outside Great Britain)[2] would provide support to the food technologists. The local NGOs in Pucallpa and Huacho would own and administer the workshops in their areas but, because of the many interested NGOs in Huancayo, no one group could be chosen to take this role and ITDG

---

[2] This is the only project considered here where one of the three international donor agencies sponsoring this research had a field office in charge of the project. In all other cases, another international NGO, local NGOs or government agencies were responsible for the implementation of the work.

proposed to retain ownership and use of the facilities. The local NGOs were to establish and maintain the contacts with the local communities. The minimum project time was to be three years beginning in July 1988 in Pucallpa and a few months later in the other two zones. ITDG was to help in the evaluation of the adoption and impact of the project technologies looking primarily at four factors:

o levels of production, distribution and consumption of basic and processed products in local, regional and national markets;
o traditional food processing methods now in use;
o changes in value added that may result from the new technologies and their socio-economic impact on those involved; and
o effectiveness of training and dissemination methods (Rios Varillas and la Cruz, 1981:9).

In Huacho, where ITDG had been involved since 1985 in the production of wines and preserves, the project would concentrate on marmalades and fruit wines (orange, strawberry, prickly pear, soursop, apple and peach), on pulping and fruit concentration, drying fruit and greens, and pickling greens. The partner agency was the Institute for Rural Education (IER). This organization, founded in Huacho at least twenty years before the project was proposed, is sponsored by the Catholic Church. Its major role is training rural farmers at its teaching farm, Primavera. The farm is run by lay people and offers short (two- to eight-week) courses to participants selected by NGOs and farmer (*campesino*) organizations. Courses include business administration, accounts, project evaluation, animal husbandry and agricultural methods. Both theoretical and practical sections are included as the commercial agricultural production of the farm is used for instruction purposes. IER also gives assistance and support to mothers' groups nationwide for distribution of foods donated by international agencies. The workshop proposed by ITDG was to be located at Primavera.[3]

In Huancayo, technologies for processing cereals, tubers, legumes, vegetables, herbs, meats, barley, maize, wheat, potatoes, peas, beans and onions were to be explored. In particular, the use of low-cost equipment, such as peelers and driers, would be studied, and pasta and toasting prototypes would be developed. The major partner organization in Huancayo was Educational Services for Promotion and Rural Support (SEPAR). SEPAR concentrates on rural education and promotion through food processing and other programmes in the Mantaro Valley (where the Huancayo project is located). It works specifically with migrants and women's groups and was

---

[3] The Center for Rural Studies and Project Assistance (CAPER) was added to IER as a partner in the coastal zone shortly after the project began. CAPER supported development projects for rural farmers in Huacho. Its association with the project was short lived.

the only local NGO with specific experience in targeting women's issues as part of its regular development activities. In 1993, the Peru Food Processing Project was one of four SEPAR projects for rural women.[4]

In Pucallpa, a major problem was food spoilage, so the principal area of work was to be the conservation of maize and rice, cassava, beans and various fruits as well as meat and fish. In Callarias (a Shipibo community on the edge of the river), the project proposed to develop a simple solar drying system for maize, cassava, beans and rice and help develop storage containers and water filtration systems. A food processing workshop was to be established in Yarinacocha (ten kilometres from Pucallpa), as part of a school already being established by the local partner organization, where hot air dryers would be developed for grains, tubers, vegetables, fruit and fish. In addition, machinery for peeling, milling and toasting using available energy would be developed, and flour manufacture from yucca, maize and beans would be studied, as would dried soup production from a mixture of flours, meat and fish. The partner organization in Pucallpa was the Center for Research and Promotion of the Amazon (CIPA). CIPA had had ten years' experience in the Amazon zone at the outset of this project. Its work involved helping Indians to strengthen their communities through learning about their legal rights, existing government services, and possible lines of communication with each other. CIPA also gave technical assistance in their economic activities, had health programmes and produced publications on themes of use to their daily lives. CIPA worked in four areas of the region, but the ITDG project proposed to work in only one, at a riverine village (Callarias) and the CIPA teaching farm. (At the latter, short courses in legal and administrative matters, communications and crop production were given.)

Because the major food producers in Peru were women, ITDG approached UNIFEM for project funding. The overall project budget proposed was $264 210. Once the project was approved, in fact, UNIFEM provided these funds and later added an additional $14 000 for gender training and participant seminars. ITDG later added $98 000 from a grant received from the ODA.

Facing the field

Work on the project was begun in 1988. During the first year, the three food processing centres were established and equipped with appropriate material for the testing and demonstration of prototype equipment for

---

[4] The Center for Farmer Services (CESCA) is a small NGO which concentrates on food processing programmes. It is cited in the ITDG proposal for training rural women in chamomile and pasta production in Huancayo town but it was only briefly associated with the project.

processing local foods. This included the development of new equipment, some based on designs provided by ITDG/UK through its other global projects, as well as the repair and adaptation of existing implements. Training in technologies was directed primarily at representatives of groups selected by the partner NGOs. These were economic groups, not cooperatives, formed from local common-interest associations such as mothers' centres, churches, a migrant society or an indigenous group. All members were women between the ages of 28 and 45, except in two units. The trained representatives, in turn, provided training to other members of their groups which formed themselves into productive units. In addition to training, the workshops provided services for the groups to utilize the equipment for their enterprise activity on a trial basis. This enabled the groups to test the economic competitiveness of the technologies and gain the necessary experience before making substantial investments in equipment purchases.

The production units, ranging from four to fifteen members each, are the 'enterprises' of this project. In the majority of cases, the individual members were both owners and workers of the productive unit and received full benefits. Such units would elect one person to be leader in charge of administration and book-keeping. In two cases, however, the productive units were solely income-generating units of the parent association; in these units, participants were workers only and had no ownership rights.

Typical investment costs for an enterprise were modest and ranged from $300 to 4000, with the average being $1500. The net profit for the two association groups was returned to the association. In the other nine units, profits were distributed three ways: one-third to the parent association (often for rent), one-third for re-investment in the business, and one-third divided among productive unit members. There were two variations on profit sharing: in Huacho, members divided produce and individually marketed it to increase individual profits; in Huancayo, groups sold collectively through established selling points and divided the profit. Most production unit members worked part time and earned $10 to 20 a month for approximately 16 hours of work.

In the first year, eight production units had been trained in a wide range of productive technologies including the processing of fruit wines, crystallized fruits, candies, dried potatoes, dried fish, soap and cassava bread. All used this training in the enterprises which they established.[5] Despite this evident progress in technology development and dissemination, other observers found problems with the project. A Junior Technical Officer at UNIFEM submitted a report at the end of 1989, which emphasized that

---

[5] This was the observation of one member of the ITDG/Peru staff and of the evaluator, F. Verdera.

the Food Processing Project never elaborated on the criteria for the selection of farm families or women's groups. These were being chosen by the NGOs on the basis of who they knew or contacts they had. Further, it lacked any formulation of what the women's situation was in the three project zones, or what their economic roles were. Nor were the staff of the local NGOs trained to work with rural women and specific women's problems. Without being able to understand the women's issues, she contended, it would be difficult if not impossible to integrate the pilot experiments into a wider community (Spada, 1989).

Given UNIFEM's focus, this report signalled a serious concern. UNIFEM staff in New York and those delegated to the UNDP office in Lima attempted to integrate women's issues into the programme but, owing to a lack of adequate personnel in Peru, could not accomplish this task. They proposed to the ITDG Lima office that the women's centre and NGO, Flora Tristan, be engaged to help re-orient the project efforts by giving training workshops to the staff. Initially ITDG refused as they saw Flora Tristan more as a feminist research centre than an establishment for field action or training. In addition, ITDG did not feel that they should finance this (perhaps unnecessary) undertaking and were not sure of the reaction of their local partners to the idea of such training (Spada, 1989). Eventually, however, UNIFEM funded training workshops for the ITDG Lima staff as well as the staff of the local NGOs and the specific project personnel.[6] These workshops sensitized staff members to the specific problems facing women in the economy and society in general. They gave substantial instruction in the theory, history and legitimacy of the women's movement. There was, however, no follow-up to these workshops and some participants complained that they were too theoretical and did not directly address the problem of how to work with local women.[7] UNIFEM also supported three seminars for project participants from different zones, which allowed them to discuss what had happened to them, what they had learned and what their general concerns and interest were.

Two major additional external problems severely hampered the Peru Food Processing Project. The first of these involved the activities of the Shining Path in the project zones. Violence threatened not only project staff and prevented ITDG or UNIFEM personnel from travelling freely to the project (Spada, 1989), but struck directly at project participants. In one case, the women's group had prepared a large supply of crystallized fruit for a special order and it was completely destroyed by a guerrilla attack (UNIFEM, 1990). CESCA, which had been one of the partner agencies in

---

[6] The training cost $8000. See UNIFEM (1991b).
[7] See UNIFEM 1990. Staff also criticized their training in interviews.

Huancayo, closed its offices and withdrew from the project because of terrorism. In Huacho, CAPER and SEPA buildings were attacked, forcing CAPER to restrict its rural coverage. Eventually, in 1991, CAPER closed its offices and discontinued its activities because of the violence. IER had to cancel one of its last workshops because of an attack on its local office in the final project year. In Pucallpa, CIPA restricted all extension activities and concentrated on the two centres at the farm and in the town (UNIFEM, 1989). Even then, CIPA did not escape the effects of the guerrillas; two leaders of an affiliated group were assassinated and equipment belonging to two other productive units was destroyed by guerrilla bombings.

Cholera erupted in the northern coastal areas of Peru in early 1991. The spread of the disease, exacerbated by poor sanitation and contaminated water in many regions, reached epidemic levels by mid-1991. Over 2300 deaths and more than 231,000 cases were recorded by mid-July. This created substantial hardships for the project participants, particularly in the Huacho zone where the incidence of the disease was higher. As a result of the outbreak, more attention had to be given to hygiene and sanitation in the preparation of locally retailed foodstuffs. Even with these precautions, productive units realized fewer sales as the public in general avoided purchases of prepared foodstuffs.

Ultimately, despite the difficulties, some successes were registered and numerous groups were reached. Eleven productive enterprises were started by the project, one of which is primarily owned by IER. Eighty-one jobs were created by the new production units. (More may have been created through the training of other women's groups but there is no record of these) (UNIFEM, 1991c). There had been no full baseline study at the beginning of the project although simple feasibility studies were conducted on different food processing products in the three zones. ITDG and local staff produced annual reports as the project progressed. ITDG/Peru staff continued all their activities. Various UNIFEM personnel worked on the project although they were unable to make field visits because of the danger of terrorism. A final report was submitted to ITDG at the end of 1991 based on a ten-day field study. UNIFEM provided additional funds for an independent project evaluation which was completed in March 1992.

The women's report

In 1993, a survey was administered to a sample of thirty-five women who had participated in one of the three sectors of the project. A control sample of thirty women who had not been involved in the project were also given questionnaires. Women in both samples were primarily from the suburbs, although a quarter of both groups came from the city. They were aged

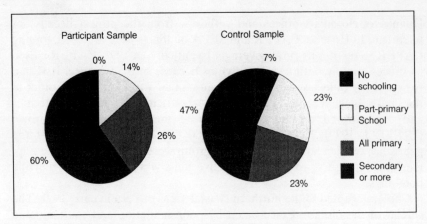

Figure 2.1 *Education: Participant and Control Samples*

twenty-one to fifty, falling primarily in the category of thirty-one to fifty.[8] Most women in both groups were not heads of their households. Consistent with women in general in Peru, most of the surveyed women had gone to school. A larger proportion of those who had been in the project had had more education than women who had not, but this is consistent with the slightly older profile of the participant group.

Eighty per cent of both groups declared themselves Catholic while the rest were Christians of another denomination. The overwhelming majority in both samples was married. Consistent with their slightly older profile, the participants had more children than the women not in the project. More of the women who had not been in the project saw themselves as housewives than did the others, but the majority of both groups were roughly divided into three categories: primarily involved in agriculture, in small enterprises or in some other unspecified but income-producing occupation. Most of both groups of women were above the poverty level for Peru, but poor. Slightly more of those not in the project fell in both a higher, and below-poverty, class.[9]

Economic impact is certainly a major criterion for project success. Most women stated that the project did not have an impact on what they did in terms of economic activities; in fact eighty-seven per cent said the project had no impact. Further, eighty-nine per cent said the project had not helped them acquire any assets. Sixty-nine per cent said the project had not helped them to acquire a larger income.

---

[8] A significantly larger group in the control sample was in the 21 to 30-year old category. The list of statistically significant relationships among variables is included at the end of this chapter.

[9] The ranking used was: upper = US$600 or more, middle = 350 to 600, low = 150 to 349, below = less than 150.

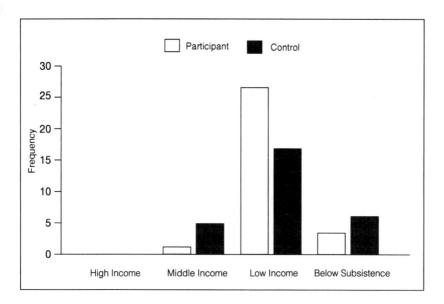

Figure 2.2 *Family Income*

In regard to family circumstances, the project was also perceived as not having had a major influence on their lives. Consistent with there being little increase in income, most women reported spending what they did earn in the same way as they had previously; in fact, women not in the project were more likely to report that they had changed how they spent their money over the period of time that the project was in operation than women participants. Most women spent what they earned on food primarily, and secondarily on clothing and education for their children. There was no significant difference in the access of women within or outside the project to luxury foods (such as beef) although the control women were slightly more likely to report eating such foods often. But many women who had been in the project (44 per cent) did think the project had a positive impact on the diet of their family because it provided them with information about better food and improved diet. Most women in the project sample (53 per cent) were likely to say that the project had improved the education chances of their children. But, when the level of education of the children of women in the thirty-one to fifty-year old group was compared through an ANOVA test, there were no significant differences in the actual education of children reported by women in both samples.

The project did not seriously affect family decision-making. In both cases a majority of women, either alone or in consultation, decided how to use their income, indicating a relatively high degree of autonomy. The same pattern emerges for the decision on what work the woman does or where the family lives. Again, most women made the decision either on their own or in consultation with no difference between the two groups. Not surprisingly, virtually all the women from the project (87 per cent) said it had no effect on the decision-making process in their family. But these women did not completely endorse the notion that the project had not affected them or their outlook on themselves and their role in society. Most of them said the project had given them greater self-confidence or made them feel better generally about themselves and their economic future.

The project did appear to have an impact on women's use of time. In this dimension, significant differences appear between the two samples. For instance, significantly more women who had been in the project said that, after the project period, they spent less time on their household and children's needs, a result which is viewed as positive here (a developmental impact) since it means more working time for the women, but has some obvious possible negative implications for the family. Significantly more women with project experience also said they spent less time on leisure and rest. In addition, an overwhelming majority of the participants also said they had more time for their own education and development than they had before. Women who had been in the project acknowledged in general that the project had caused them to alter their use of time in these ways.

Interestingly enough, almost half of the women said they wanted even more time for work and more than fifty per cent said they wanted more time to do what they wanted. In this regard, however, they are not unlike the non-project women, among whom more than fifty per cent said outright that they wished to have more time for their work. Given that one of the goals of improving the lives of women is to lessen their work and give them more leisure, it is interesting that the women themselves did not share this ambition. They wanted more time to work; it was the only way they perceived of changing their lives. Perhaps, it was a way of gaining increased indpendence, but in any case it was their only means of improving the living conditions of their family. In the short run, they are undoubtedly right – that only by more work would they increase their income – although, instead of decreasing the burden they carry, this impact of the project and their own ambition can only result in increasing their load.

Overall, the women were glad they had taken part in the project. Two-thirds of them praised the project for offering some combination of training, credit, technical support and back-up work. Most of them (81 per cent), however, could find things which they would have liked improved. Some said the project should have been better planned and should provide longer agency support. Others would have liked more and better training, while others wished the project had done feasibility and market studies to help their enterprises. Almost a third wished the project could have been improved in some or all of these areas, and that it had provided them access to credit.

These results indicate that the project had some developmental impacts on the women who participated, perhaps more than might have been anticipated by those who reviewed the programme in the final evaluation in 1992. The women who participated did claim they had more self-confidence and felt generally better about their economic future because of their experience. This suggests that just being supported in group activities and receiving training in the use of new technologies may open to women the vision of being able to do something for themselves, of actually being able to change the bleak future which otherwise they have to accept as being all there is. However, in terms of actual or concrete impacts on their lives, very few can be discerned from this study. Women are not generally empowered in their decision-making, they have not acquired a larger income or more assets and they certainly have not gained more leisure. What they did get, because they were given training at least in technology use, was more time for self-development, which obviously built their self-esteem and confidence. They gained additional information about improving their diet which many women felt was a help to them. These things are important, but again, these gains were not in the context of overall concrete improvements in their lives.

Nor did their families reflect much improvement in their situation. In fact, asked what impact the project had had on their families, most women said it had none. Twelve per cent did say it provided more income for the family, but a quarter of the women said, if it had done anything to the family, it was to reduce the time the woman had to give her children or her husband. Thus, the overwhelming majority said it either had no effect or hurt the family. Surprisingly, although there are reports of verbal complaints of various male family members that their wives could no longer devote enough time to their children, most men were not completely negative in the survey results. They liked the idea of their wives learning how to get a greater income.

In sum, the results suggest that the Peru Food Processing Project had a slight positive developmental impact because it made women feel more confident and more able to do something to improve their situation. However, in terms of most of our measures, it had no impact. It did not change women's access to income, acquisition of assets or their living conditions in any other significant way.

One other line may be explored, however, to see what the actual effect of the more intensive approach – mobilizing women and working directly with them to backstop their enterprises and help them overcome the problems they encounter in use of the technologies, management of accounts, marketing, etc. – might have had in contrast to the hands-off, simple technology development and training approach largely typical of this project. In our analysis of the project, we indicated that SEPAR, the agency working in Huancayo, was different from the other local agencies because it did have an orientation to mobilizing women, doing realistic feasibility studies on their proposed enterprises, helping train them to manage their enterprises and giving them other technical support when necessary. Thus, looking at women participants from this region and comparing them to participants from other regions may provide some evidence of the degree to which another, more involved, approach to local women might have had a greater developmental impact.

Marisela Benavides, the local researcher, reports a large difference from the interviews she did in the three zones. In Huancayo, for example, training from SEPAR in the San Martin production unit had a significant impact on the women participants' general ability to change their own environment. These women have obtained financing to get electricity and pure drinking water to their community. They see themselves as 'dynamizing agents' in their own communities. The women in the production units are very proud of what they are doing, and are eager to improve and expand their work. They recognize certain problems which they have, such as a lack of defined marketing strategy, but they hope to overcome such obstacles. Furthermore, apart from the income from the project, the training they have given to other women in their community has been very valuable. Many women have had training in technical food processing and in nutrition, and some have gone on to purchase, or otherwise obtain, the tools for food processing enterprises which they plan to set up in the future. Ms Benavides found the experience of the women involved in the production units in this zone of the project very striking:

> It was very stimulating to talk with some of the *empresarias*, not only in regard to their own expectations and self images, but also on the total role they play in their communities: seeking basic services, solving immediate problems and taking definitive leadership roles.

She adds that the Huancayo women involved in the project (in the production units) appreciated the training they received. It enabled them to feed their families better, to handle their own finances more skillfully, 'not to be so timid' and to improve their relationship with their husbands. Those women to whom she talked, who had not been in the project, were envious of the participants. They wanted the economic security of the project women; they wanted a chance to be involved themselves. The husbands of the participants had a more mixed reaction. Some were very supportive but others resented the project and said it took too much time and meant that the children did not get sufficient care. However, Marisela Benavides believed that this negative attitude could and did change with time as the women's production units became more successful.

In contrast, it was difficult to find a reaction or ascertain the impact of the project in Pucallpa. Here the major enterprise was the production of bread, but interviews suggested that this was done irregularly. There was no established market for the product. The women, who were extremely shy of talking to outsiders, had no idea of what income they received from participating in the project, if any. Nor did they know about their own family income. Unlike the situation in Huancayo, the only training or extension work had been in the preparation of bread. According to Ms Benavides, the solely technical direction of the training made a major difference between the results of the project in Pucallpa and that which occurred in Huancayo. Research in Pucallpa, however, was hampered by the inability of the women to converse directly with Marisela Benavides as they did not speak Spanish.

In Huacho, the third and final area, only one of the project enterprises, los Cipreses, is still operating (1993) and producing marmalades, jams and other fruit products. This is seen as an important achievement because women's groups in this region had not previously been production units or earned income for their families. Los Cipreses is doing that despite economic difficulties. One of the problems here, similar to Pucallpa, was the lack of anything but technical training as well as the absence of a marketing study or marketing support. In addition, in Huacho, more than anywhere else, Marisela Benavides recorded a response that, because of fear of terrorists, women could not continue their work. Perhaps because of the lack of any kind of community awareness or social organizational training, those who received short courses did not go on to set up small enterprises. But the food processing training had an impact, in this case especially on the staff of the local NGOs. Two of the local NGO promoters who were trained went on to do further useful work with that training. One of these went on to teach the techniques she learned at a local school and has even set up a small enterprise with parents and teachers of the school. Another is working with children in the local Sacred Heart parish and has trained them to produce sweets.

Evidence from the local researcher's work strongly suggests the importance of adopting the approach of mobilizing, training in management and accounting, and technical support after the enterprise is begun, if the developmental goals set in the beginning of this section are to be attained. Looking now at the survey data, it can be seen that her analysis is born out quantitatively. When the sample of participants is divided into those from Huancayo, as opposed to the rest, certain interesting and significant differences do emerge. ANOVA and T tests show that, for those from Huancayo, there was a significantly greater probability of them saying that the project had an impact on their economic activities. In addition, Huancayo women were significantly more likely to say they now spent less time on home and family and on leisure and resting than they had before the project, and more time on their own education and self-development, while the others were much more likely to say their use of time had not changed. Huancayo women were more likely to say the project had improved the family diet while the others said it did not affect theirs. In addition, the women of Huancayo were much more likely to say the project increased their decision-making power. Women of Huancayo were more likely to decide on the education of their boys and on where the family lived. These differences from the rest of the sample may be attributed to the traditionally greater family authority of Huancayo women (see above), but the changes they report over the project period can not be disregarded. Only the women of Huancayo reported an actual change in many significant areas over the time of the project, such as gaining more decision-making power over their work than previously. They were more likely than women from Huacho or Pucallpa to say that the project increased their income and otherwise improved their lives. Thus, despite needed caveats about possible alternative explanations for the distinct difference between Huancayo women and other participants, our results support the hypothesis that the greater the attention paid to mobilization and training of women and to the economic sustainability of their projects, the more likely the developmental impact of a project will be achieved.

Conclusion

The overall assessment of this project is complicated. If it is looked at in terms of its own goals adopted at the outset then it must be judged at least in part a success. Among its major achievements in terms of what it set out to do were the development of appropriate technologies for the processing of thirty-two products. In addition, thirteen machines for the processing of fruits, grains, tubers and beans had been introduced. Furthermore, information materials were developed to help other groups and individuals interested in the specific technologies. These included five technology handbooks on dried potatoes, fruit wine, fruit candy, vinegar, and (processed) corn. These guides included a summary of the process and costs. A video was prepared on the potential of food processing as an income-generating activity. Not scheduled but highly desirable was a partnership with the private sector in the manufacture of new equipment which also resulted from the project. Using ITDG designs, one local workshop in the Huancayo zone, CIDEMETAL, has been particularly successful in opening new markets and further disseminating the new technologies through its promotional efforts. The workshop began production with start-up capital and initial orders from ITDG. These overall successes of the Peru Food Processing Project were particularly important because they demonstrated the potentiality of the development and transfer of technologies to small enterprises in a country where state institutions which work on technology usually work with medium and large-sized concerns. Thus, according to the ITDG report, this project was a significant counter-example to the current trend to support large enterprises and traditional technologies (Begazo, Caceres and Verdera, 1992:4).

Not all the evaluations of the project were uniformly laudatory, however. The project evaluation report pointed out some severe flaws. In the first place, although technological objectives had been achieved (in successfully identifying and developing food processing technologies), the project was not able to reach as many people in as wide an area as had been intended. Nor was it able to set up a procedure to adequately support women's small enterprises established with these technologies through business training, access to credit etc. Even the project staff recognized the failing, stating that there had been little progress in 'two key factors which go beyond food technologies: the business administration aspect and the inclusion of specialized work in aspects of gender' (Begazo, Caceres and Verdera, 1992:3). Because the project was not successful in these areas, it did not reach a point where its work could be usefully transferred to other organizations, including mainstream institutions, as a prototype of what could be accomplished. This had been of primary interest to UNIFEM (UNIFEM, 1990).

ITDG was engaged in developing food-processing technologies and making these available to local groups. This it did extremely well, achieving considerable success in the range and extent of food processing innovation and adaptation. This is a monosectoral, or 'simple', approach as the emphasis was purely on technology development and demonstration. When UNIFEM began to closely observe what was happening under the auspices of the local organizations, it perceived the lack of focus on women and women's community development with the results stated above. UNIFEM then insisted on, and carried out, workshops to train the staff. These had an impact; among reports from project staff is the recurring theme that this was a learning experience for local institutions and for ITDG. But these workshops were not sufficient to redress the balance. There was no follow-up and little practical content to the workshops. Part of this is due to the lack of personnel with experience in gender issues and in working with women within a framework of experience of successful women's projects. Such people, appropriately trained for working with food processing technologies, are not so easily available. The local partner institutions did not have them (SEPAR was the only partner with any gender awareness background). ITDG had only begun to train its Peru staff in this direction.

The collaboration of ITDG and UNIFEM in this instance offers an interesting example of a project for women's microenterprise development. Both agencies stood to learn from the contrasting approach to development used by the other, and both changed their perspectives somewhat as a result. The project evaluation report cites a long list of things which should be done differently in another project. What seems most important from this list, however, is that a new project by these two partners contain a carefully designed framework for working with women and agreeing at the outset on:

1) the community development/social awareness goals, how they are to be achieved and how much time and effort are to be devoted to this;
2) how much time and effort should be given to training and support of small enterprises (and what exactly would be done in this regard); and
3) to developing, demonstrating and backstopping technologies for food processing or other production.

The conclusion from this analysis is that the criticisms of the Peru Food Processing Project in the external evaluation have merit if the goals include a positive developmental impact on the women who participated in the project. This project, in so far as it did not establish from the outset mobilization and management training of local women and assistance with credit, marketing and management as a chief element in its requirements of the collaborating agencies, was doomed to fail in achieving any major impacts on the women's lives. Only where the local agency did have this approach were these desired ends met. What this ITDG/UNIFEM project did was good in its own right, but limited in terms of its results for the women who participated – except where SEPAR was involved. Perhaps, the women's increased confidence and group experience using the technologies may some day be translated into more concrete changes in their economic activities and income and their lives, but there is no proof of this type of result as yet.

Chapter Two: Survey Analysis Results in Peru: Significant Findings

1) Sample Type (women participant, control) and Group Membership (belongs to economic group, does not)
   Chi sq. = 42.162, p. = .0001, Phi = .805

2) Sample Type (women participant, control) and Change in Use of Time (no change, spends less time on family, household)
   Chi sq. = 7.437, p. = .0243, Cramer's V = .341

3) Sample Type (women participant, control) and Change in Use of Time (no change, spends less time on leisure and rest)
   Chi sq. = 4.808, p. = .0904, Cramer's V = .278

4) Sample Type (women participant, control) and Change in Use of Time (no change, spends more time on self education and development)
   Chi sq. = 11.891, p. = .0026, Cramer's V = .503

5) Sample Type (Huancayo participants versus other participants) and Impact of Project on Economic Activities (had no impact, had positive impact – more income, more activities etc.)
   ANOVA unpaired t value = -2.533, p = .0239

6) Sample Type (Huancayo participants versus other participants) and Impact of Project on Time for Leisure and Rest (no impact, spends less time on l. & r.)
   ANOVA unpaired t value -4.392, p = .0001

7) Sample Type (Huancayo participants versus other participants) and Impact of Project on Time for Self-Development and Education (no impact, spends more time on self-development and education)
   ANOVA unpaired t value -4.249, p. = .0002

8) Sample Type (Huancayo participants versus other participants) and Impact of Project on Family Diet (no impact, family diet improved)
   ANOVA unpaired t value -6.644, p. = .0001

9) Sample Type (Huancayo participants versus other participants) and Impact of Project on Respondent's Ability to decide on Whether She Works and What She Does (no change over project period, had more decision power)
   ANOVA unpaired t value -1.979, p. = .0562

10) Sample Type (Huancayo participants versus other participants) and Impact of Project on Respondent's Life (No impact, increased income)
    ANOVA unpaired t value -5.108, p. = .0001)

11) Sample Type (Huancayo participants versus other participants) and Impact of Project on Respondent's Life (no impact, increased income and otherwise benefited respondent)
    ANOVA unpaired t = -1.867, p. = .0708

12) Sample Type (Huancayo participants versus other participants) and Impact of Project on Respondent's Decision-making Power (no impact, increased decision-making power)
    ANOVA unpaired t = -5.011, p. = .0001

CHAPTER THREE
Honduras – Cashew Nut and Fruit Production

The state and the nation

Honduras has a population of 5.1 million people living in a land area of 112,000 km^2. Average per capita income is $590 annually, much lower in the southern rural zone where this project is located (World Bank, 1992:218). Primarily an agrarian country with agriculture generating about 23 per cent of GDP, (World Bank, 1992:222) 75 per cent of exports and 55 per cent of employment in 1992 (Europa, 1992:1315–21; IMF 1992:392–95), its chief commodities for export have traditionally been bananas and coffee. These have, however, suffered massive swings in prices owing to the instability of the world market. As a result, the economy of Honduras has been extremely unstable over the last ten years. Only 44 per cent of the population lives in a large town or city. Southern Honduras, where the Cashew Project is located, is more rural and more dependent on agriculture than the rest of the country. Approximately 70 per cent of the population of this region gets its livelihood from agricultural work. The south also has been characterized by far higher unemployment than the rest of the country, with a rural unemployment rate of 16.06 per cent in 1988 compared to the national average of 8.18 per cent (GOH, 1988).

During the 1980s, the Honduran economy was strongly affected by a massive infusion of US foreign assistance in the form of military and economic support. Indeed, the foreign aid package in 1987 was equivalent to 22 per cent of the federal budget of Honduras. This assistance, ensuing as a matter of policy from the Nicaraguan conflict, artificially inflated the economy at a time when the real market forces were producing an economic slowdown. This economic instability was somewhat offset by the inflows of capital in the form of US assistance, so that the Honduran economy managed to maintain a low inflation rate throughout the 1980s. This occurred despite the downturn in agriculture and the accumulation of large foreign debts resulting from the costs of accommodating the Nicaraguan Contra rebels. In 1990, shortly after the conclusion of the Nicaraguan conflict, the Honduran government carried out a devaluation of the Honduran lempira. Inflation increased sharply over the next two years but has since begun to stabilize. (The average annual rate for 1987–91 was 13 per cent, but by 1992, the rate was only 8.8 per cent.)

The Government of Honduras has made attempts to support poor rural farmers and small enterprise production through legislation and government action, although the results have been limited. Land in Honduras had been controlled by large landowners and the predominant style of agriculture, appropriate for banana and coffee cultivation, was plantation-style: large commercial farms on which poor peasants worked as labourers. By the 1960s, peasant farmers had begun to organize into various production, marketing, self-help and political groups. They began to place increasing pressure on the government for land reform. The government responded with the land reform acts of 1972 and 1975. But the reform was short-lived and reached few *campesinos* (rural dwellers). By 1982, only 55 000 families (14 per cent of the rural population), had been granted occupation of 600 000 acres, or 9 per cent of Honduras's farmland (Yudelman, 1987:35–36). These laws of Agrarian Reform and the later Law for the Modernization and Development of Agriculture were, however, first steps towards recognizing the rights of small farmers.

In 1990 the Honduran government began a credit programme, the Honduran Fund for Social Investment (FHIS). In fact, this fund offered very limited amounts of credit to rural microenterprises because it primarily supported other industrial, commercial and service activities. In 1993, a new programme called the Program for Credit, Training and Technical Assistance for the Rural Microenterprise (PROCATMER) was established with funds from the Honduran government and from external donors. However, it has not had an interest in the type of production sponsored by the Honduran Cashew Project.

Government recognition of the needs of small farmers and of the importance of small enterprise development, against the backdrop of a population which has been politically organized and active for many years, is an important factor, despite the paucity of actual state resources directed to this area. Other factors are less positive. For one, the country is still recovering from its indirect involvement in the Nicaraguan crisis, in particular, the southern zone where the cashew project is located. The instability and insecurity resulting from this situation are diminishing but have not disappeared. Relatively high unemployment, inflation and currency devaluation all combine to make the economic situation for the small farmer or entrepreneur extremely difficult. In 1970, 65 per cent of all Honduran households were below the poverty line (Catanzarite, 1992:71). The situation has improved since that time but a plurality of Honduran households still fall in the lower economic, below-poverty category.

Women in Honduras

As in Peru, women in Honduras have a second-class role in the economy and society of the country. In regard to education and literacy, however,

Honduran women have made considerable progress. In addition to formal education, the government of Honduras has embarked on several adult education programmes which women are more likely to attend than formal schools. The results of this are indicated in the literacy figures. Fewer women are illiterate than men among people aged 15 to 24. Out of people aged 25 and over, women are more likely to be illiterate indicating that this trend toward educating women has been in more pronounced in recent years. Women are only slightly less likely to go to primary school than their male counterparts (most boys and girls do get this schooling). In regard to secondary school, women are much more likely to go. It is only at the college or university level that men receive more schooling (UNIFEM, 1991:52). The increasing education of women is a positive indication of the change in the role women play in Honduras. Neither the state nor society, nor the church, opposes women's schooling, and women have clearly found reasonable access.

Women's relative disadvantage to men, however, shows in the economic position they occupy. Only 21 per cent of Honduran women are recorded as economically active by state statistics as opposed to 86 per cent of Honduran men. As of 1990, women were only 20 per cent of the active labour force (UNIFEM, 1991:106). This statistic is questionable. It states that women's labour force participation has dramatically decreased since 1950, but women are certainly not working less now than they were in that earlier period. This may be primarily a result of a change in the criteria as to who is seen as 'active' by the government census takers since rural women are not inclined to be full-time labourers because of their other responsibilities and thus may have been eliminated from the 'active' category (Catanzarite, 1992:73). In any case, women *are* a small minority of those in wage-paid formal jobs in Honduras at present.

Women, particularly urban women, are very active in the informal sector. Recent female migrants to the city are often domestic workers, but those who have stayed for any length of time are more likely to move into informal sector activities. The state has records of domestic service employment but not of the percentage of women in the informal sector (Catanzarite, 1992:73–5).

In Honduras, the male head of household has the greatest family authority, especially in the rural areas. Where families do own land, it is registered in the name of the male. Nor is it common for a women to have a bank account apart from the family account controlled by the male head of household. Certain factors, however, combine to give women a degree of independence. The first of these is the trend in Central America away from dual parent households, especially among the poor. The majority of women are still in long-term monogamous relationships, either sanctified by the church or common law, but a distinct and growing minority of women are not married (but have children) and are the heads of their own households

(Catanzarite, 1992:70). They have authority over all decisions although, of course, no contributions from the male parent to lighten their responsibility for family well-being. Secondly, women are expected to be involved in economic activities even in households where there is a male present. Agricultural and other tasks are divided by gender. Subsistence crops grown around the house and food processing at home, for example, are ordinarily a woman's task, while men control the production and marketing of cash crops. Women normally decide on the use of their own revenues or salary, when they are lucky enough to have one (which they are likely to use for family needs in any case).

There are numerous women's groups throughout the country organized for production, marketing and other mutual aid purposes. Since the sixties at least, the Catholic Church and other organizations have encouraged and supported these. It is not uncommon for women to join such a group for economic activities. Membership opens possibilities otherwise not available, such as obtaining credit, having a bank account, etc. The Honduran Organization of Rural Women (FEHMUC), for example, has more than 5000 members from thirteen of the eighteen departments of the country. Many are single mothers. FEHMUC's programme is directed to consciousness-raising and organization, health and nutrition, agriculture, crafts and clothing production. In each area of interest, there is a complex set of programmes funded largely by outside donors (Yudelman, 1992:36–45).

The Cashew Processing Project takes place in one zone in the Department of Choluteca and the towns of Namasigue and El Triunfo. This is the southernmost zone bordering on El Salvador and Nicaragua and was traditionally one of the poorest and most backward areas of an already severely underdeveloped country. The area has suffered from weak agricultural yields caused by ecological devastations from years of timbering, the expansion of the cattle industry and climatic fluctuations and bouts with drought. The agricultural productivity of small-scale farmers has suffered; most farmers harvest only one-third the amount of subsistence crops produced by their counterparts in other regions. Short falls in food production were met with food aid, provided through 'Food for Work' programmes, until 1989.

Given the dismal outlook for agriculture and the high rate of unemployment, both men and women have emigrated in search of better opportunities. A USAID study estimated that, since 1974, emigration from the southern region has averaged 1.3 per cent annually. According to this same study, in the early 1980s when the Cashew Project began, 41 per cent of all southern region families did not meet minimum subsistence standards. Moreover, nearly 65 per cent of children under the age of five experienced stunted growth, and the region's average infant mortality rate of 99 deaths per 1000 live births was substantially worse than the national average of 61.9 deaths (USAID, n.d.).

The Honduran Cashew Project 1981–8

During the early 1970s, The Honduran Government introduced a number of donor-financed programmes to help develop agricultural alternatives for the rural population. One such programme included the promotion of cashew trees as a reforestation alternative producing a potential cash crop. With support from the Inter-American Development Bank (IDB), more than 100 peasant co-operatives received financing for the planting of cashew trees. The IDB loan was to be repaid once the trees came into production. Shortly after the programme initiative, however, government assistance decreased and the farmers did not receive the requisite technical training in planting, cultivation, processing or marketing of what was to them a non-traditional crop. Furthermore, around this same period, the Choluteca region was increasingly affected by the Nicaraguan civil war as many Contra rebels used the area for staging raids into their country. Although the Contra presence affected only a few of the cashew growers directly, the concomitant infusion of military and economic assistance inflated the Honduran economy, resulting in an overvalued currency. This rise in the lempira's exchange rate adversely affected external trade promotion efforts for cashews (as well as other Honduran exports) by rendering Honduran prices non-competitive. In this set of circumstances, faced with a poor market situation and lacking training, it was not surprising that cashew nuts were being left to rot and cashew trees were being burned for firewood.

The Cashew Project developed in response to a request in 1980 to Pueblo-to-People for marketing assistance from one of the Choluteca cashew-grower groups. By this point, the IDB-funded government programme was no longer active and cashew growers were seeking help to learn how to process and market cashew products. Pueblo-to-People today is a major Alternative Trading Organization (ATO) with over eighty-five projects in nine countries, but in 1980 it was quite new (See Annex to Chapter Three, p. 66). As stated above, the Cashew Project in Honduras was its first country programme. Although Pueblo-to-People did not have experience in cashew processing, its founder, Dr Dan Salcedo, did know a number of food scientists who might be able to provide the processing assistance the farmers needed. Dr Salcedo approached Barrie Axtell, a former colleague and a food scientist, who was working with the Institute of Nutrition for Central America and Panama (INCAP) in Guatemala. Axtell (who later joined ITDG and was instrumental in the Peru Food Processing Project) assumed the central role in researching methods for cashew nut processing. He had developed a method for processing the 'false fruit' or apple of the cashew which became a major part of the project and a source of employment for women. ITDG, then, provided technological information and support for the project but had no role in implementing it.

The Pueblo-to-People Cashew Project had nine basic activities:

1) Development and dissemination of a cashew processing technology appropriate to the skills and resources of the growers' groups in Choluteca.
2) Establishment of a quality control system. Pueblo-to-People provided hands-on training and frequent monitoring to upgrade processor skills and ensure hygienic standards.
3) Establishment of post-harvest systems: methods for storing cashew nuts and fruits to diminish post-harvest losses.
4) Development of packing systems using US-made plastic bags.
5) Guaranteed purchase of all processed products. Purchases were guaranteed even when quality or production levels were of low standards.
6) Identification of markets. Pueblo-to-People targeted special market niches such as alternative-style health food stores and progressive consumers in the US. Pueblo-to-People also developed a sale catalogue in which the nuts and dried fruits were featured.
7) Arrangement of transport for processed nuts and fruits from the field to the market.
8) Establishment of an accounting system. Pueblo-to-People designed an accounting system that records available working capital credit, field purchases, sales, outstanding loans and profits.
9) Hands-on training of the staff of a local NGO, the Asociacion Proyecto del Pueblo (APDP). Pueblo-to-People involved its local staff directly in developing the methodology of production and the process of export marketing. Staff became familiar with permits required, product packaging, freight forwarding, etc. APDP became an NGO in its own right in 1988. APDP works only with peasant organizations, in particular those involved in the agrarian reform sector, in areas of technical and marketing assistance. While APDP does promote its services, it does not specifically develop projects but only responds to direct requests for assistance. With specific reference to the Cashew Project, cashew producers and processors have a contractual relationship with APDP and are fully responsible for their individual projects. They participate in marketing decisions through their representation in ADIM, the cashew growers association. The ADIM board, in turn, has input into APDP through selection of APDP management staff.

An additional and essential component of the project was provided by HIVOS (The Humanist Institute for Development Co-operation), a Dutch charity, after the APDP took control of the project. HIVOS gave money for group start-up costs and for a revolving fund to allow the purchase of raw materials and to help pay for promotional activities. HIVOS supplied money for the salary of the ADIM (the producers' group) full-time coordinator, who played a crucial role in intermediating between APDP and ADIM. HIVOS also funded training for staff members of APDP.

Unlike most agency projects, this one began with no pre-determined time period or fixed operating budget. Nor was there a baseline study or a full feasibility study, although the IDB had carried out a study on cashew nut production in Central America (IIES, 1990:Chs 1, 2). A major part of the operational costs of the project were initially borne by Pueblo-to-People and thereafter by HIVOS. The revolving fund which it provided, as of 1993, was valued at 186 000 lempiras or $32 632.[1] Training for APDP staff in NGO management, purchase of commodities, and recurrent costs are estimated by APDP to equal $120 000. ITDG gave forty days of technical assistance which is an additional cost for which we do not have exact figures.

Most project costs were low as Pueblo-to-People depends on self-financing projects, although there are significant costs which have been borne by Pueblo-to-People as part of its general overhead, of which we have no estimates. Later, after the cashew project had become a substantial business, further donor assistance was required. Other than HIVOS and ITDG, Pueblo-to-People obtained assistance from the US Peace Corps which provided five volunteers over the life of the project who gave critical help in book-keeping and general management. The Post-Harvest Institute (USA), Fine Dried Fruit Association and several private groups gave unspecified amounts of additional support. In 1988, Pueblo-to-People transferred its project responsibilities to APDP, the local NGO it had established to carry on the trading and market development role of Pueblo-to-People. The project (from the point of view of the original international agency support) could be said to have ended in that year, although Pueblo-to-People continues to maintain contact with the Cashew Project as one of the primary marketing outlets for APDP and still provides technical assistance when needed.

Events and challenges

Initially, the operations of the Cashew Project were very modest as the organization lacked a funding base and required its projects to be self-financing. The approach adopted in the Honduras programme focused on the development of export products by small-scale producers for the purpose of establishing self-sustaining agro-processing groups. Barry Axtell, at Salcedo's request, worked on the processing of the false fruit of the cashew. This was a new process and was first marketed internationally by Pueblo-to-People in this project. At the same time, the US-based staff of Pueblo-to-People was conducting market research and developing market strategies[2].

---

[1] The exchange rate in May 1993 was $1 = 5.7 lempiras.
[2] Dan Salcedo and his wife, Marijke Velzeboer, sold cashews on the streets of Houston to ascertain the market reaction.

Table 3.1 Net Sales of Honduran Cashew Products through Pueblo-to-People ($) (P-P, 1993)

Year	Cashew Nuts	Cashew Fruit
1985	$7 139	—
1986	*9 792	*2 000
1987	24 394	3 676
1988	16 552	3 465
1989	19 442	5 155
1990	21 769	3 826
1991**	17 913	3 826
1992	12 690	1 660
1993	***3 900	***1 000

* Estimate   ** Reflects mealy-bug infestation   *** First quarter figures only

From this simple beginning, the Cashew Project grew to include the processing of both the cashew fruit and the cashew nut and established a significant market for these products. By 1985, the Cashew Project represented a significant part of Pueblo-to-People's overall global business, with over ten percent of the organization's income deriving from cashew mail orders. Sales of cashew products through Pueblo-to-People are shown in Table 3.1.

One significant achievement of the project was the increasing inclusion of women in the processing of cashew nuts, and their dominance in the production of the cashew fruit. In its initial stage, the Cashew Project had been a response to a request from men's agricultural production groups and had aimed at disseminating improved technologies to them. Though the men took the primary role in harvesting, management and marketing, their wives and other family members were, from the outset, given the primary task of processing the nuts. These women were hired and (in theory) paid as daily workers. Over the time of the project, the participation of men waned as most men's groups dwindled in size; a few completely withdrew from processing and shifted all responsibilities to their wives.

When the project introduced the processing of cashew fruit, this was solely directed to women's groups. Men's groups showed little enthusiasm for the dried fruit production because it was only a part-time and seasonal activity with a limited, and initially untested, specialty market.[3] Cashew nut processing, on the other hand, was more lucrative as a full-time, year-round activity. However, men have begun to lose interest even in the nut production. In recent years, other crops in the Choluteca zone have become

---

[3] The cashew fruit is naturally bitter and unpleasant tasting and had had no appeal until the ITDG research established a method of reducing the unpleasant taste and sweetening the product. Thus, unlike the nut which had had a traditional market, the market for the fruit candy had to be created.

increasingly profitable. There has been development in the zone of commercial shrimp farms which directly employ over 7000 people and indirectly provide work for 100,000. The zone has thus become a stronger centre of economic activity. Although women are the major (direct) employees of the shrimp enterprise (60–70 per cent), men are being drawn out of the less profitable cashew business to this activity as well as other economic alternatives, thus making room for women to become increasingly central in cashew nut and cashew fruit processing. Women, of course, have been drawn to the other economic alternatives now open to them, but the persistent lack of sufficient opportunities means that there are still many women for whom cashew nuts are the only source of income, particularly since cashew nut production is done at home, and cashew fruit production is only a part-time and seasonal activity.

Presently, APDP is working in seven communities. It works with five men's groups, but only one is still involved in all aspects of processing. In the other male-controlled groups, women who work are often the spouses of the male producers. More recently, two women's groups have engaged in the processing of nuts on their own. Since none of the women's groups have trees in cultivation, they purchase the nuts from their husbands or other cashew tree owners. Women in these groups tend to be older or single women heads of households; younger women with spouses tend to prefer the part-time fruit drying option. There are six women's groups processing cashew fruit associated with APDP. All women's groups have a membership ranging from five to fourteen women. The overall processing activities involve 76 men and 83 women directly.

The majority of the groups are informal and have no legal constitution; two have formed co-operatives and one has formed an incorporated business. All members of each type of group are considered 'owners' of the business with 'shares' in the group's assets. The requirements for membership (in addition to being nominated) include a minimum investment, no outstanding debts, and good references. All group members share in the profits at the end of the year. While most groups divide profits among members according to the amount of time worked, some extend the same dividends to salaried workers. Each group elects a governing body to administer overall affairs and maintain finances. Most groups (including the women's groups) employ salaried help. There is no distinction between trained and untrained workers. Most of those working with nut processing earn at least 15 lempiras ($2.63) a day, while those working with fruit drying earn between 10 and 12 lempiras a day ($1.80–2.25). Nut processing enterprises work nine to ten months a year; fruit drying groups work four months per year.

In 1988, Pueblo-to-People had transferred responsibility to APDP, thus ending its direct involvement in the project. APDP continues activities including the arrangement of packing, transport of goods from the field to the market, monitoring of quality control, price negotiation, arrangement of payment and liaison to the producers through the producers' group, ADIM, liaison to Pueblo-to-People and the identification of new markets. Pueblo-to-People continues to have an important role, however, in three major areas. It provides access to technical advice, its staff acting as intermediaries in researching and identifying technical resources. For example, Honduran cashew producers have suffered a major infestation of Indian mealy bugs. Although this plague affected the amount of cashews that could be marketed, Pueblo-to-People continued to purchase all of the exported crop while it sought assistance in dealing with the problem. In 1988, losses from this infestation amounted to over $50 000. In 1990, Pueblo-to-People, with the assistance of another ATO, Cultural Survival, contacted the Post Harvest Institute for Perishables which finally identified the problem as one of inadequate storage. Based on this advice, post-harvest handling practices were changed. Pueblo-to-People estimates that in 1990 alone, a single technical adviser spent over 15 per cent of his time on this infestation problem. The costs for his services were absorbed by general Pueblo-to-People overheads.

Pueblo-to-People still provides access to improved processing methods through its consultations with other ATOs and private sector firms about ways to improve cashew processing methods. In collaboration with Cultural Survival, Fine Dried Foods International recently visited the project site. Its report included recommendations on such issues as processing and wetting agents, moth control measures in processing and packaging, mould control of the finished products, solar drier construction and design, packaging and improved cooking methods (Fine Dried Foods, 1991–2).

Finally, Pueblo-to-People still provides APDP referrals from other ATOs such as Cultural Survival and other NGOs such as Oxfam in order to open new markets for the Honduran cashew products. Recently, Pueblo-to-People helped research the legal requirements for the project's cashew products being named a 'fair trader label producer'. This certification would facilitate the entry of Honduran cashew products into preferential markets of organizations such as Oxfam that promote the sale of products by Third World Producers.

Despite the evident success of this initially small project in creating a market and stimulating and supporting small-scale production of cashew products among women and men, certain severe problems were encountered during the process of project implementation, such as the infestation of mealy bugs, which led to significant losses. The maintenance of an edge in processing technology was also a problem. After the initial work of ITDG there was no further technology adaptation or follow-up until the visit of an expert from Fine Dried Foods International in 1992. He identified a number of basic problems such as improper use of the solar driers, most of which did not face south, improper loading of trays resulting in drier damage, limited maintenance and poor construction of driers. He found minor design faults such as undersized vents which caused condensation and inhibited dehydration; and driers which were located too close to the ground, forcing women to remove the dried fruit at night to prevent animals from eating them (Fine Dried Foods, 1991–2). In regard to the nuts, the Honduran project faced a basic problem because the quality of the nuts produced was low. Only 45 per cent of the final quantity of processed nuts could be classified as whole and thus meet higher quality standards. In 1993 APDP staff received training from the Friedrich Ebert Foundation in improved processing methods which it intended to introduce to the processing groups.

Finally, an improvement in the Honduran economy was a problem for the ultimate success of cashew processing enterprises. As a result of the changing opportunities available, men started shifting their activities to more commercially viable crops. Furthermore, APDP points out that local *processing* has recently decreased because many cashew growers find it easier and more profitable to sell raw nuts to local and regional traders from Guatemala and El Salvador than to sell to local processing groups. With the return of political stability following the end of the Nicaraguan conflict, regional trade in cashews has increased as foreign importers have demanded higher quality nuts than the Honduran cashew project usually produces. The recent establishment of a local medium-scale processing plant in Choluteca, by the Federacion de Productores y Exportadores de Honduras (FPX) – through a private sector firm, Maranons del Sur (MARSUR) – threatens the market for the Cashew Project's nuts further (This project is an ATI project which also has USAID funding.).[4]

The women's record

Several factors must be kept in mind in analysing the data on the actual impact of this project. In the first place, it started twelve years ago so the impacts should be more thoroughly internalized than in shorter projects. Secondly, and of great importance, the project began as a response to a *men's* group of producers. Women were gradually incorporated into the project and have moved into more central positions, not only in cashew fruit production but also in the processing of the nut, as men have found higher-paying options. Men, however, were the ones first supported by the

---

[4] The ATI Choluteca project was established because of the falling production of cashew nuts due to the low quality of the cashews produced (making the product non-competitive on the world market), the low rate of productivity of the nuts and the low prices received by farmers (giving them no incentives to replant trees or ability to buy needed fertilizers, high-quality seeds for replanting etc.) (ATI, 1991).

the project; women worked for them in the household on the production process and, in some cases, continue to do so. Men remain both the owners of the cashew trees and the heads of household in the traditional Honduras system. Secondly, although the training was not evenly distributed and women's mobilization was not systematically undertaken, women were given training in the use of the appropriate technologies and the management of their enterprises, and they were assisted with credit and marketing.

The survey in Honduras was administered to thirty-five women who participated in one of the cashew groups and to a control sample of thirty women, also in the southern zone, who did not participate. Most of the women in both groups lived in a purely rural area although a significantly larger group of the women who had been in the project came from towns. Most of the women were between twenty-one and fifty years old and most were married and not the heads of their household. The majority reported their husbands were farmers (78 per cent of the participants and 59 per cent of the control women). More than a third of both samples had not been to school but the remainder in both groups had been to some or all of primary school. Virtually all of the women were Catholic with a small minority declaring themselves Christian but not Catholic. Most women had between four and ten living children.

The first significant difference between the samples comes in their perception of their occupation. All of the participant women saw themselves as working in a small enterprise at home while the control sample, although also involved in productive work, considered themselves housewives.[5] This may or may not indicate that the project women saw themselves with slightly higher status than did the uninvolved women who, in fact, took care of animals (cows) or poultry as a primary non-domestic activity. The project women also were engaged in economic activities other than their cashew production. More than eighty per cent cared for poultry, for example. For more than half of the women not in the project, the small income they earned from their animal husbandry, poultry or vegetable garden was their only income. Clearly, the cashew enterprise was a significant source of income for the women participating in the project although they also received a small income from these other economic activities.

Significantly more project women were members of a women's group than were uninvolved women, 97 per cent of whom were not. This difference is not surprising as the project only worked with women's groups which they helped organize. Women's groups were not as common in this zone as they are in other parts of Honduras.

---

[5] A List of Chi square, ANOVA and T Test significant results is included in the Annex to Chapter Three.

Most project participants had joined in the early stages of the project and most said they joined because they expected to achieve a larger income. In fact, these women did think the project had helped them – 77 per cent said that it had led to a change in their economic activities and an increase in their income. A few women did not think it changed the activities they engaged in but did believe it increased their income – 93 per cent said it increased their income. The non-project women were about equally likely to have assets as the project women and virtually none of either group of women earned income from their assets. Interestingly, however, those participant women who had land or a house, had acquired that asset during the project period and believe that the project had been the cause of their good fortune. Finally, without full baseline data on either the participant or the control women, it is hard to draw firm conclusions, but, it may be significant that, although the participants did not prove to have become wealthy as a result of their involvement in the project, a significantly *smaller* proportion did fall in the below-subsistence category than the control women.[6]

Both groups of women were from poor rural or semi-rural areas, and both were poor overall. Therefore, it is not surprising that their expenditure patterns are very similar. Most women spent the income they earned primarily on food and only secondarily on other things such as clothes or medical expenses. Project women reported that this was the same

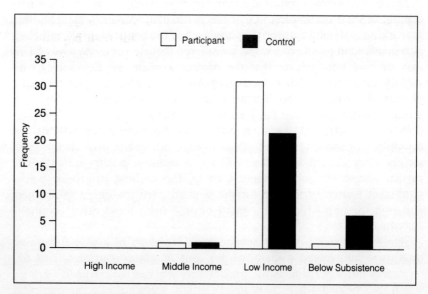

Figure 3.1 *Income Levels*

---

[6] The standard is set here with four groups; upper, middle, low (= 3 000 to 3 500 lempiras), below.

expenditure pattern which had always existed and that the project had had little impact on it.[7] In regard to actual family circumstances, most women in both groups reported that their families did not eat expensive foods (such as beef), but participants did perceive that the project had had an impact on their family diet. Seventy-seven per cent said the project allowed them to eat better food than they had before and, within this group, seventy-eight per cent said it was because of the extra income they now received. Ninety-one per cent of the participants said the project allowed their children to have more education than they would otherwise have had, although, when the actual levels of education of the participants' children were compared to the control children (through ANOVA tests), there was no significant pattern of difference.

Regarding the authority of the women to make major decisions affecting them and their families, the results indicate an interesting trend. Project women, for example, are significantly more likely to say they make their own decision about how to use their income (83 per cent versus 44 per cent), while the other women are more likely to say they consult their husbands on this matter. There is no difference between the groups of women in regard to who decides whether they work and what they do – in both cases a plurality of women said they made their own decision, while slightly more than a third said they consulted their husbands and twenty per cent said their husbands made the decision. Women from the project were significantly more likely to say they decided what the family would eat. As far as deciding on the level of schooling for both boy and girl children, again the project women were more likely than other women to say they made this decision. Thus, it appears there is some support for concluding that women from the project had more authority in more domains than their uninvolved sisters, but certainly not consistently and significantly across all decisions.

Nor did the project women see their decision-making power as a result of the project. More than eighty per cent said the project had not changed how decisions were made in their family and an even higher percentage responded that their families had not changed their decision-making patterns over the period of the project. However, these women appeared to mean that the project had not changed their overall pattern in all family decisions. They did think the project had an impact on their own feelings of self-worth and self-confidence. Asked in their overall summary of the project whether it had an impact generally on their decision-making power, seventy-

---

[7] The questions pertaining to possible changes in economic activities, family conditions, decision-making and use of time over the period of the project were supposed to be administered both to the participants and the control groups. To the latter, the interviewer was supposed to merely ask about changes over the time period (in this case twelve years). In the case of Honduras, this was not done so we have no comparative measure of the changes in lifestyle of the control versus the participant group.

one per cent said yes. Asked how it had affected their self-confidence and outlook for the future, fifty-two per cent said it made them optimistic about their economic prospects, thirty-three per cent said it gave them confidence in their own abilities and only three per cent said it had no impact.

Although no comparison is possible between the particpant and control samples in Honduras in regard to their use of time, women from the project exhibit a pattern quite similar to the Peruvian project women. A majority said they spent less time now than they had before the project caring for their children or their household, while only a third said they had not changed their use of time. Fifty-one per cent said they spent less time resting or in leisure activities than before, although a few women said they spent more time doing so. Three-quarters of the women said they spent more time on their own education and self-development than they had before while a handful said nothing had changed in this regard. Almost all the women said that the project had been the reason for the changes in how they spent their time. Most women (fifty-seven per cent) said they would like to have more time for their work. This is significantly different from women who had not been in the project who did not wish to change their use of time.

Not surprisingly, the women who had been in it liked the project. Most said it gave them training. Forty-seven per cent said it helped them obtain credit and, as stated above, ninety-four per cent said it allowed them a greater income and/or more assets and/or more work for the family. One hundred per cent said their family benefited through increased income, more education for the children or more work. Most of the participants said their husbands appreciated the project because of their increased earning ability, and virtually none of them reported negative attitudes on the part of their spouses. Indeed the sample of husbands supports this contention – one hundred per cent liked the project because it provided more income, more education or more work opportunities for the family. In striking contrast to the Peruvian sample, eighty per cent of the Honduran women participants said they saw no need for change in the project. Eighty-eight per cent saw multiple project strengths from agency support, through training, credit, new technologies, marketing or some combination.

In sum, these results suggest that the Honduras project had a strong positive impact on the women who participated in it. This impact included creating feelings of self-worth among the women and changing their time use as well as their appreciation of that use. From the women's perspective, what was most important was that the project increased their income and improved their expectations of future income. Project families also had a better diet than previously. Project women were much more likely to control their own income than non-project women. What the project apparently did not do was strongly support women in enabling them to recognize and assert their own equality. Remembering again that this was initially a men's project and that women worked first for their husbands in processing cashews, we find indications from these results that men's authority remains largely dominant in the family. In fact, seventy-seven per cent of the participant women reported assisting their husband's work, most of these in processing the cashew, others in tending the plants or other activities. Despite the initial condition of the project that women should be paid for their work, seventy per cent of the women who worked for their husbands said they were not paid. Most (60 per cent) reported that they were assisting their husbands just as they had been before the project started.

The women participants had certainly been mobilized to some degree by their participation. While not equal in the family, they had begun to play a new role through their access to income, the expenditure of which they determined. Thus, they were now contributing to family expenses and seen as more important within the family structure. Their membership in women's groups, required for joining the project, provided them with a solidarity experience which gave them an understanding of what women's co-operative activities could achieve with training, credit and some outside technical support – without any dependence on male aid. These were not small accomplishments.

The comments made by the women reflect their feelings:[8]

> 'Now we are united and work in an organized fashion . . . we get training from other women who have been trained by the project, and help when we need it . . .'
> 'We can do anything . . .'
> 'The project makes me feel better because now I am more secure about my economic future . . .'
> '. . . I can buy uniforms and school supplies for my children . . .'
> 'We all live better now because of our increased income . . .'
> 'I can buy better food for my family now . . .'
> 'Now I can pay for my house and I have no fear of my future . . .'

Thus, although women still have not achieved parity in their households, this study suggests a much stronger and more positive set of impacts on the Honduran than the Peruvian women except for those in the Huancayo subsample.

Conclusion

The Honduran Project has lasted for more than twelve years through varying conditions and changing responses from both the market and local

---

[8] The quotations here are taken from the participant questionnaires and translated from Spanish.

participants. Several significant accomplishments of this effort should be noted. First, the project was able to give small-scale cashew producers the expertise to process marketable cashew nuts and dried fruit, and connected them with both a local and an international market. This action is the more important because these were resource-poor peasants with low product yields, marginal lands and few competitive crops. Commercial agriculture in the zone, dominated by cotton, coffee and cattle in large production units, controlled most of the arable land. Few of the commercial operations had any interest or resources to devote to transferring new techniques to small producers, or developing arrangements for contract farming. This project cut through directly to the needs of this marginalized group. Presently, the project is responsible for the creation of 200 full-time jobs and 60 part-time jobs. Over the life of the project, substantially more jobs were created but we have no record of exactly how many.

The second major accomplishment was certainly the development and transfer of technologies for processing cashew nuts and cashew fruit. Prior to the project, farmers had no knowledge of how to process cashews. Nor was knowledge available, even in the agricultural research institutions in the country, to make the processing of cashew fruit seem a viable commercial activity. Currently, both products are produced by local farmers and exported and, in the case of the nuts, locally consumed.

Thirdly, the project developed fair trade links between local producers and the international market and established a niche for the cashew products. This was the result of Dr Salcedo and his wife's market surveys, test-marketing of nuts and fruit, and identification of potential buyers in the United States. Local producers were taught to negotiate independently and manage business relationships in the wider commercial market.

Fourthly, the cashew project was able to establish an income-generating opportunity for women, first as labourers in the nut production, then directly through the cashew fruit, and finally in nut production as women began to take a more central role in the nut processing. In addition, the project's success with women led to its replication by other NGOs who established similar projects for women in the region.

Finally, the project led to the creation of the Cashew Nuts Producers Association (ADIM) and the APDP. The latter is an NGO which has been able to take on many of the functions of its progenitor, Pueblo-to-People, in Honduras; the Cashew Project is only one of its activities.

Reviews of the Honduran Cashew Project, although acknowledging these achievements, have been quite critical of a number of failings. In the first place, the Honduran Cashew Project has only worked with a few producers groups – seven men's groups and seven women's groups – reaching after twelve years only approximately 200 men and women producers directly. Recently APDP has tried to expand the project beyond the original groups of producers to other groups, but has been unable to do so because new groups are only willing to join if offered the same type of subsidy as was available to the older groups. Pueblo-to-People had funded the construction of the processing plants for the original groups at a cost of about $640 per group.[9] APDP could not continue to offer this kind of support and new groups refused to join the programme – they felt they could not participate – without it. In other words, the programme could not self-replicate because it had been too dependent on agency support. This dependency went a step further to the essential question of marketing. Pueblo-to-People had discovered, indeed created, a market for cashew nuts and cashew fruit in the United States but APDP has been unable to extend this work significantly. APDP and the project are still dependent on Pueblo-to-People for major foreign marketing and for covering production problems through subsidy from the latter, as happened during the mealy bug infestation.

Secondly, the project had concentrated on research and dissemination of appropriate technology and marketing of the product, but not on organization or training of the producers. This poses a severe problem to the continuation and/or expansion of the project. Over the life of the project, the group enterprises have been troubled by such management issues as lack of business planning, group administration, leadership, communication and financial management. This has contributed to the desertion of the project by many of the original male participants (IIES, 1990). The Honduran project did not have the resources for training (other than in technology use) but relied on programmes sponsored by the government within the agrarian reform sector to provide training in administration and organization to individual groups. The 'generic, disparate and often arbitrary' training thus available was not adequate to the needs of the producers. Some received little training, others repeated. One co-operative, for example, had received thirteen courses in general organization and administration from six different institutions (Lobo, 1993). In the absence of adequate training, the producers' groups could not operate efficiently or effectively.

A third major problem was in the very technology on which the success of the project was based. ITDG, at Pueblo-to-People's request, had developed and tested a technology in nut and fruit processing which was appropriate to the skills and needs of the Honduran producers. This was a significant breakthrough. But there was little follow-up work on continuing or newly developing problems with the technology, or further possible adaptations or improvements of that technology or its use by the local people. The Institute for Socioeconomic Research faulted the project for not moving beyond artisanal levels of processing which were no longer

---

[9] This money came from HIVOS.

economical (IIES, 1990). A lack of standardization and a failure to upgrade technologies was noted in a USAID/ATI feasibility study of the cashew nut industry (ATI, 1991). The study by the Fine Dried Foods International expert confirmed these criticisms. In the processing of cashew fruit, for example, he noted that no changes had been made in the process originally developed by ITDG six years earlier. And, he found a number of serious problems and proposed a series of simple solutions which would enhance the production, appearance, and taste of the finished products. In addition, the technology adopted had serious health risks resulting from unbearably high temperatures and poor ventilation (Fine Dried Foods, 1991-2).

A last criticism is in the area of project assistance to poor rural women. The project did open income opportunities for women deliberately, did work with women directly, and did support their role in the organization of producers (ADIM). Given the lack of opportunities for rural women in Honduras, this is significant. But the project had no defined and developed plan for helping women establish their position. Thus, its efforts to work with women were hampered by opposition from men who resisted their inclusion. For example, when the project tried to offer courses for women, spouses would not let their wives attend, a reluctance which APDP attributes to the fear that women might become too independent, especially if they developed their own financial resources. APDP has sponsored two courses on the rights of women and is looking for additional funding to do more in this area, but the project suffered from this not having been adequately addressed earlier.

Like the Peru Food Processing Project, the Honduras Cashew Project has been most strongly criticized for not achieving goals which it had not set for itself at the outset. Pueblo-to-People never expected to do anything but support technology research and dissemination and, by expanding and creating markets for cashew products, stimulate the growth of a successful rural producers' enterprise. This it did extraordinarily well, especially at the outset. This agency had neither the staff nor the expertise to do more than this. In any case, like the Peru project, this may be seen as a learning experience, at least for its original sponsor, Pueblo-to-People and the new local NGO, APDP.

Finally, from the perspective of the developmental impacts on the women who joined the project, the results suggest these were considerable and positive. Women's incomes, overall well-being and sense of self were improved because of their participation. In addition, participation in women's groups gave the Honduran women a sense of solidarity and capacity for achieving desired goals through their own mutual efforts. Technology and marketing were the keynotes of the success of this project but it was the combined package of direct attention to women through training, not only in technology use but in management, credit responsibility, marketing, etc., direct assistance to their enterprises in terms of technical assistance and through marketing assistance, and mobilization training in the women's groups, which paid off. All of these strategies together had highly desirable developmental impacts, even though these stopped short of giving women a sense of their own equality in the household or in their businesses. The clear benefits of the project for the women, however, are undercut by the prognosis reached here of possible future failure for the cashew enterprises as presently organized and with their current technology, and by the impossibility of project replication. What will become of these women – will they be able to take advantage of the new income opportunities growing up in their region and capitalize on their recent training and greater self confidence? – this remains to be seen. The latter is a seminal question to be answered before any final evaluation of this project can be made.

Annex – Chapter Three

Pueblo-to-People is a non-governmental organization based in the United States. The organization began in 1979 when Dan Salcedo, a US citizen with a doctorate in operations research, was in Guatemala working for the United Nations. (He had also worked for INCAP.) The purpose of the organization is to help the poor by selling their crafts in the United States so that they can make a larger profit. Pueblo-to-People today buys from approximately 85 production groups in nine Latin American countries: Mexico, Guatemala, El Salvador, Honduras, Nicaragua, Colombia, Peru, Bolivia and Brazil. It is one of a growing number of Alternative Trade Organizations (ATOs) which use international trade as a way of getting assistance to the poor in developing countries (P-P, 1993).

Pueblo-to-People's approach is to respond to requests from production groups for help with their processing and marketing. It also sells imported products through a mail order catalogue and at many fairs, festivals and conferences. It has a retail store in Houston, Texas. Pueblo-to-People's staff includes twenty full-time people, four part timers and four volunteers. They are chosen for their grassroots development experience in Latin America. Many have graduate degrees. Pueblo-to-People's Vice President is Marijke Velzeboer, Dan Salcedo's wife and a Dutch citizen born in Colombia. She was associated with the founding of Pueblo-to-People, and has a doctorate in public health and a specific interest in the organization of projects for women. The importance of the roles of Drs Salcedo and Velzeboer in Pueblo-to-People, and in the Honduran project specifically, can not be overstressed.

Pueblo-to-People had a gross income of three million dollars for the fiscal year ending June 30, 1992. Approximately 40 to 45 per cent of this figure goes to the producers (and also covers freight, duties and brokers' costs in the US) (P-P, 1993). This project was a central part of Pueblo-to-People's Honduran country programme).

Chapter Three: Statistically Significant Survey Analysis Results in Honduras

1) Sample type (participant, control) and occupation (housewife, animal husbandry, small enterprise)
   Chi sq = 40.63, p = .0001, Cramer's V = .791

2) Sample type (participant, control) and membership in economic group (belongs, does not)
   Chi sq. = 57.201, p. = .0001, Phi = .938

3) Sample type (participant, control) and ownership of other assets (owns, does not)
   Chi sq. = 4.992, p. = .0255, Phi = .281

4) Sample type (participant, control) and Family Economic Level (poor, below subsistence)
   Chi sq. = 4.204, p. = .0403, phi = .254

5) Sample type (participant, control) and decision on use of income (wife decides, wife and husband decide)
   Chi sq. = 9.909, p. = .0016, Phi = .406

6) Sample type (participant, control) and decision on where to live (wife decides, husband decides, husband and wife decide)
   Chi sq. = 17.385, p. = .0002, Cramer's V = .517

7) Sample type (participant, control) and decision on family diet (wife decides, husband decides, husband and wife decide)
   Chi sq. = 8.112, p. = .0173, Cramer's V = .353

CHAPTER FOUR
Guatemala – Wool Production and Processing Project

Economy and politics

Guatemala has a population of 9 200 000 living in a land area of 109 000 km². It is less urbanized than Honduras, with only 39 per cent of its population living in towns or cities. Agriculture provides only 26 per cent of its GDP, while services provide the major portion or 55 per cent (World Bank, 1992:218, 222). Nonetheless, Guatemala has a small, solidly developed manufacturing and industrial base. Its economy, however, is characterized by a series of severe contrasts. For example, there is a sharp disparity between small subsistence farmers and larger commercial holders. The country has one of the most unequal structures of land ownership and income distribution in Latin America. That disparity, along with the dominance and legacy of former military rule, have led to years of violence and political instability which in turn have slowed industrial development.

The Wool Production and Processing Project was initiated in 1985, shortly after the installation of the first civilian government in Guatemala in over sixteen years. With the return of an elected government, tourism – an important factor in the demand for artisan wool products – increased by almost fifty per cent in the year before the project began. By the end of the project, it had reached over 500 000 tourists a year (Europa, 1992: 1261–72). Increased tourism not only provided a local outlet for handicrafts, but more importantly, stimulated a demand for the export of quality artisan products. Despite this positive trend, other negative economic factors continue to plague the country. Among these is inflation which, although nowhere near as serious as in Peru, has been a problem for businesses and especially small producers with small profit margins. Thus, the inflation rate for 1987 was 12 per cent and by 1991 was up to 33 per cent, slightly lower than the previous year but still a matter of some concern.

As in Honduras and Peru, the informal sector of the economy has grown rapidly over the last twenty or more years. The urban informal sector experienced particularly rapid growth during the period of economic and political crisis of the 1980s. In 1986, estimates of the number of microenterprises in metropolitan Guatemala City alone ranged from 45 000 to 125 515 (Revere, 1990:89). Informal sector activities are also common in the countryside, although microenterprises are fewer. Recognizing the importance of this sector, the Guatemalan government, shortly after the transition to civilian rule, adopted a policy of supporting microenterprise activities through credit, training and technical assistance. The two major programmes are the Urban Microenterprise Program (SIMMER) and the program of Financial Intermediation for Comprehensive Development (IMF). SIMMER has provided loans and training to microentrepreneurs since February 1988, with the goal of training 25 000 in the first three years and creating 75 000 jobs. IMF, also established in 1988, is a rural programme aiming to strengthen local economies by promoting the formation of producers' associations whose members benefit from large-scale purchases and sales. Both programmes are designed to work through NGOs and use their informal sector experience (Revere, 1990:89–105).

Certain drawbacks have emerged from this programme. For one thing, NGOs which collaborate with SIMMER or IMF must use the latter's methodology of operation rather than their own tried strategies. In addition, inexperience has led to weaknesses in the SIMMER and IMF programmes, such as too low interest rates, inadequate training and poor monitoring and evaluation systems. Further, NGOs felt pressure from SIMMER and IMF for performance goals but received little support (Revere, 1990:93–6). As a result, many experienced NGOs do not participate in these government programmes and there has been a growth of independent private sector programmes to serve the small-scale enterprise sector. Since 1988, the number of NGOs working with microenterprises and small businesses in rural and urban areas of Guatemala has grown to more than twenty. Among these is the Foundation for the Development of Socioeconomic Programs (FUNDAP), the local agency in charge of the Wool Production and Processing Project.

The government's supportive stance toward small-scale enterprise development facilitated the development of programmes such as the project studied here in two critical ways. First, it generated strong support from the international donor community, expanding access to resources available only through government-to-government funding. Second, the government's decision to work through NGOs instead of a new ministry demonstrated the importance of these organizations in working with small enterprises, and resulted in opportunities for these agencies to work with the donor community. This government programme, with its focus on SIMMER and IMF, has been criticized for not taking a broader view to change the laws and public policies which discriminate against the informal sector (Revere, 1990:93). Nonetheless, it was a generally positive background for the development of the wool production sector.

Guatemalan women

Women are disadvantaged in Guatemala by common practices as yet unreformed by education or law. Many changes are, of course, taking place; the area of education is one of the most striking. Among older women, 67 per cent are still illiterate but, among women 25 or younger, that figure has fallen to 52 per cent. (Rural areas are, of course, worse than urban ones; 75 per cent of rural Guatemalan women are illiterate.) For every hundred boys in primary school there are eighty-two girls, and for every hundred boys in secondary school, eighty-three are girls (UNIFEM, 1991:52; Marina Delgado, 1992:32). This is a lower rate of female education relative to boys than in Honduras but it does indicate substantial progress. Women are also entering the labour market. In Guatemala women are 16 per cent of those in the formal sector. They are still under-represented in most major areas of employment, comprising only 19 per cent of the administrative and managerial workers, and 14 per cent of labourers, but they have established certain niches – women are 93 per cent of clerical, sales and service workers (UNIFEM, 1991:106). Unfortunately, women still occupy the lowest status and least paid positions. The most common paid employment for a Guatemalan woman, for example, is as a domestic servant. In recent years there has been a rapid increase in women's factory employment as well.

As in Honduras, the husband is the head of household in Guatemala. The wife is unlikely to have independent status through such things as a bank account or land registered in her name, although she is able to decide on the use of the money she earns as salary or from her other economic activities. Her choice of expenditures, however, is determined by her family's needs. In addition, as in Honduras, there is a relatively large number of women who are heads of household and who are therefore in charge of all aspects of their lives and of those dependent on them. This is in part because of a high male emigration rate from rural areas to the cities and from Guatemala to the United States. In addition, monogamous relationships tend not to endure (as in Honduras) and thus women frequently live alone with their children. Also, many men were killed in the civil war. Almost half of the women heading households are widows. Fifteen per cent of all families have a woman head of household (Marina Delgado, 1992:35). (This figure is higher in Guatemala City) (Catanzarite, 1992:70).

Owing to an intensive study of women conducted by Luz Marina Delgado in 1992 (as a mid-project evaluation conducted for the Wool Production and Processing Project), considerable information is available on the structure of the lives of women in the project zone. The project was based in the northern mountainous region of Quetzaltenango, about a four-hour drive from the capital, Guatemala City. Indigenous Mayan farmers populate the highlands and cultivate small, isolated farm plots. The project region is one of the poorest areas of the country with most of the population engaged in subsistence farming and living in bare poverty. In the higher altitudes where the wool is produced, families derive their main income from flocks of about fifty sheep. Middlemen generally buy the wool from the families at low prices and sell it in the wool market in the small town of Momostenango, the centre of the wool processing and weaving industry.

Most of the weavers live in the mountainous countryside near Momostenango where weaving, rather than sheep farming, provides the main source of family income. In these weaving households, the men of the family weave wool rugs, blankets, and clothing. The women handcard and spin wool and sometimes assist in dying the yarn needed by their husbands. Traditionally, the wool products are sold to middlemen for sale in the local economy, or in the capital for the tourist and export trade.

The area's high population density and land scarcity have forced many Indians, particularly sheep farmers, to look for work elsewhere. Delgado's study estimates that two to three out of every five families have been affected by migration (Marina Delgado, 1992:11). As a consequence of this emigration, wool supply had fallen prior to the beginning of the project, as many farmers turned their herds over to other family members to tend while they sought work elsewhere. For weavers, this meant a lack of adequate and quality wool with which to make their finished products.

The project area has long been characterized by a climate of fear and insecurity. The Quetzaltenango region has suffered from political unrest since the mid-1950s when the military assumed control in a US-backed coup. Guerrilla activity was centred in the zone. The government, in turn, instigated a long series of counterattacks, and human rights groups have accused it of persecuting, torturing and murdering civilians – particularly the indigenous Indian population. Warring rural guerrillas, left-wing urban terrorists, right-wing counter terrorists, and the military have caused local and regional upheavals and produced an overall sense of instability and insecurity throughout the region.

During the civil war period, the government engaged in repressive tactics to stifle opposition. It also, however, promoted a 'food for work' programme. Guerrilla activity intensified in response to the government's repression, escalating violence in the countryside. Over one million Mayan Indians, caught in the middle of this conflict, either fled to Mexico or migrated illegally to the United States (Europa, 1992:1261–2). Those who remained did so in an atmosphere of great mistrust of the government and most outside institutions. With the election of the civilian government in 1986, the Guatemalan government struggled to achieve national reconciliation, and some, though not all, of the violence in the rural areas diminished. The Wool Production and Processing Project was initiated in 1987 within this context of violence, unrest and extreme disturbances.

Against this backdrop of unrest, there has been little attention to, interest in, or resources for improving the lives of women. Women in the zone are in a situation of 'open disadvantage in relation to the male population' (Marina Delgado, 1992:24); this despite the relatively higher status and greater freedom that Mayan women traditionally have relative to some of the other ethnic groups in Guatemala. The subordination of women is greater in the project's poor rural zone than in Guatemala City or other more urban areas. Looking, for example, at schooling, women are less likely to receive an education here relative to men – in one school, for instance, only 9 per cent of the primary students were girls. Most (60 per cent) of the poor uneducated rural women have from six to twelve children. This large number of children affects the health and longevity of women in the zone, and limits their time and opportunity for income-generating activities.

Traditionally, Mayan women had worked in various crafts, the products of which they sold. They were traders and carried out all local commerce, but the introduction of industrially produced products and trade across trucking routes from the capital reduced the value of their products and led to the displacement of women from the trading role; men became the agents. There is still a strict division of labour by gender in the region, although, as in regard to trade, this has evolved over time and in response to outside incentives. Women process corn, care for small animals and poultry, make the food, sew cotton clothes, wash clothes and care for the children and the house. They also fetch water and grow market-garden crops for family consumption. Men plant and tend the corn, produce wool, raise large animals and build the houses. Both may be involved in the production of non-traditional plants such as potatoes. Most women who earn money do so through their production of various items from crafts to processed foods, but they receive very little, often not even covering the costs of this operation. Men receive and control the income from the sale of corn or sheep or the production of wool. Few women belong to socioeconomic groups such as co-operatives or mutual aid societies. In fact, women are not likely to be involved in community activities of any kind; more than 60 per cent are not. Most are fully occupied with their work in the family business, alone except for family members, isolated from larger society, unappreciated and with few opportunities to improve their lives (Marina Delgado, 1992).

The Wool Production and Processing Project

The project originated when a group of weavers asked FUNDAP for marketing assistance. In response, FUNDAP began a small USAID-financed project to assist in commercialization. Within six months, FUNDAP recognized that the constraints affecting the weavers were broader than just marketing and included the processes of wool production and its sale to the weavers. FUNDAP started to develop an integrated sector strategy which would attack all major barriers. With the Ministry of Agriculture and

USAID support, FUNDAP approached ATI to help in performing a project assessment, designing a revised project plan, and providing direct inputs into the implementation of the revised plan. Ultimately, in the revised plan which ATI and FUNDAP prepared jointly, the project goal was to increase the incomes of rural Guatemalan peasants involved in the wool subsector through improved production, processing and sales of export-quality artisan wool products. To achieve this, the project used a vertically integrated subsector approach based on a series of technology developments and linkages in three areas: sheep production and improved animal husbandry practices with small sheep farmers; improvements in wool carding, spinning and dyeing with weavers; and the establishment of a commercial enterprise for improved quality control and marketing of finished wool products.

The revised project plan retained the essential components of FUNDAP's original project, but significantly expanded the sheep production component to ensure the upgrading of the quality and quantity of wool produced. The enhanced project added several specific technological innovations in wool handling and processing, as well as activities in marketing and commercialization. FUNDAP was to be responsible for field implementation and administration. ATI would provide technical assistance, in particular to the wool production component, as well as management co-ordination of all project components.

The project was divided into three distinct components, each implemented in different geographic areas and supported by independent teams of local FUNDAP staff. FUNDAP maintains its central offices in Quetzaltenango, one of the larger urban centres of Guatemala. The various project activities are between one and four hours drive from Quetzaltenango. ATI and FUNDAP saw this joint effort as a particularly important project because it was one of the first in which these agencies used a subsector approach in identifying a range of interventions for small-scale producers which would enable them to establish commercially viable enterprises. ATI had a further interest in this work because its staff was eager to work with FUNDAP. ATI staff knew of FUNDAP's earlier successes and believed its approach particularly suited to ATI's own philosophy and programming.

Sheep and wool production
The team for this component worked with sheep farmers through Technical Assistance Centres (CATs) to improve breeding, animal husbandry and other technologies to ameliorate the quality and quantity of raw wool provided to artisans. Participants included sheep farmers in San Marcos, Huehuetenango, and Nenaj-Quiche. The CATs provided the primary outreach vehicle for regular contact and development work with sheep farmers. Each CAT was located in an area accessible to 50 to 100 sheep

farmers and provided farmers with common-use facilities for meeting, storing wool, and sheep pens. The project's local promoters were based at each CAT. These centres were used for all technical assistance training in animal husbandry and other improved practices and were the collection points for the wool sold and picked up for market sales. At each CAT, selected members were trained to carry out, on a cost-reimbursable basis, extension work on animal hygiene and feed production. Loans were available to farmers through the CATs for both the purchase of seeds for pasture and for equipment.

In April 1989, the project was modified to include the creation of a centralized wool-washing plant to be owned and operated by a national organization of sheep producers. The need for such a facility (in which to classify, wash and dry wool) was acknowledged after previous project attempts to upgrade traditional washing techniques had failed.

Oversight and technical assistance was provided by a FUNDAP team supervised and assisted by an internationally recognized sheep/wool production specialist, the late Dr Ian Fraser, who was contracted by ATI.

Wool processing and weaving
The FUNDAP team, led by a FUNDAP expert in artisan crafts, worked directly with individual artisans to provide hands-on assistance in the production of wool products. The early phase of the project focused on the establishment of a weavers' co-operative, COPITEM, accomplished in 1989. Throughout the project, the weavers selected representatives to work with the FUNDAP project manager as employees of COPITEM. The primary interventions provided by the project were channeled through COPITEM. In addition to training in new production techniques, interventions also included access to credit for the purchase of inputs and structural improvements in the artisan workshops. COPITEM, membership of which is open to any interested weaver, received a grant from the Belgian Foundation for Development (FODEP) for construction of a co-operative service facility. Project services such as a materials supply bank, a dyeing laboratory, a mechanical carding machine, a show room, and a training centre were established at this facility.

Commercial enterprise
The team charged with this component worked closely with the wool processing and weaving group and focused on improved access to markets through the establishment of a legally recognized commercial enterprise. This marketing unit, INNOVA, specialized in the commercialization of Guatemalan handicrafts. It provided assistance to weavers in the planning, design and market testing of new products. INNOVA also assisted with the establishment of quality control standards for finished products and with market surveys.

Costs for this project were borne by ATI (from its USAID core funding) and by USAID-Guatemala through a separate grant to FUNDAP. ATI estimates its total contribution to FUNDAP to be $309 279 and its home office costs $104 000. In addition, ATI paid Dr Fraser through a separate contract, the total of which was $91 356. The USAID-Guatemala grant to FUNDAP was $399 084.

Overcoming the obstacles

The project was a large-scale and complex undertaking with many elements to co-ordinate. ATI retained a major role in its overall management and co-ordination although FUNDAP had direct charge of project field management. Essentially, in the sheep-raising component, the team worked with local full-time farmers ranging in age from 31 to 45, and with only primary schooling. Few had had any contact with other development programmes and all were essentially focused on their own sheep raising business. These businesses were individual family based, having the male head of household as the 'owner' assisted by other family members. He received and controlled the use of the profits, although Mayan women have strong influence in their households (see discussion below). Other family members do not receive a salary for their work. The weavers are also engaged in individual family businesses, structured again with a male head or owner assisted by other unpaid family members. The weaving is done by the male head.

The project established a weavers' co-operative, COPITEM, which has 120 members and employs nine full-time staff. Benefits of membership include training, access to input stores, and sales through the co-operative store. Co-operative profits are divided between retained earnings, which are reinvested into the business, and member dividends. For an interim five-year period, the co-operative retains member equity, including all dividends. At the end of five years, the members can withdraw in proportion to what was invested on a yearly basis.

The marketing enterprise, INNOVA, was established to market all FUNDAP project commodities. INNOVA is incorporated as an association and owned by a consortium of artisanal co-operatives and FUNDAP. It employs four individuals. Dividends from its sales are given to the co-operative owners and are reinvested in individual co-operative services. Individual weavers do not receive anything directly from these sales but do enjoy improved services.

AFORAM is a veterinary pharmacy established by the project; it employs one person whose salary comes from the sales profit margin. FUNDAP initially established and purchased the inventory for this enterprise.

The project registered substantial successes because of the booming tourist business and the increased demand for wool products of high quality. By the project end, weavers' incomes had increased by at least 20 per cent. Sheep farmers' income in general had gone up even more dramatically, by about 50 per cent.[1] Income had doubled in the areas where sheep farmers have successfully reduced the mortality of their flocks. COPITEM was operating successfully as was INNOVA and AFORAM. The wool-washing business had added substantial improvements to the quality and the price of the products of the farmers – by more than a third. The weavers had learned the use of new technologies and different designs and dyes. Although these changes had not begun to spread widely through the zone, FUNDAP believes that about ten per cent of all local weavers, not just project participants, have started to change their traditional modes of production to take advantage of these new elements.

External problems along the way had slowed progress and limited the quantity and quality of the final results. Inflation, although not as serious a factor as in Peru, still threatens participants' profits. In addition, the climate of violence, suspicion and distrust remained a serious factor throughout. The project had had to eliminate one of the three regions chosen for the sheep production component, Nebaj-Quiche, for this reason. In other regions, the natural wariness of the farmers and their families, and the absence of a tradition of group projects made the work of supporting project activities very difficult. The sheep farmers were especially isolated, distrustful and resistant to working with the project. Everywhere, the farmers and weavers exhibited the same lack of faith after years of broken promises from the government.

As a result, the organizational development of groups was very difficult. The early phase of the project focused on the formation of both a weavers' co-operative and a sheep farmers' national association. It proved far easier – although still difficult – to organize the weavers. Efforts to organize the sheep farmers on a national level had little success. The successful formation of the weavers' co-operative, COPITEM, followed an initial period during which disgruntled members of a defunct weavers' group attempted to block the project. The situation was complicated by the programme manager's lack of development experience in this area. The recruitment of a new manager with more relevant skills improved FUNDAP's outreach efforts. The new manager's efforts to defuse the situation and incorporate a more participatory approach saw immediate results in the increased involvement of weavers in the nascent co-operative group.

Another related difficulty in organizing weavers was the legacy of the 'food for work' programmes which had established a 'welfare' mentality. According to the project manager for the sheep component, one group of

---

[1] If inflation is taken into account, these percentages are lower. However, the comparison between those farmers in the project and those outside indicates a substantial improvement for participants not enjoyed by others (Loarca, 1990).

weavers expressed no interest in new training if food donations were not forthcoming. Another difficulty revolved around the traditional fixed relationships between weavers and middlemen. These relationships, often developed over generations and within family circles, required that weavers commit their sales to certain traders who, in turn, provided them with credit, transport, and ready markets. Many weavers were hesitant to sever these established relationships to join a new organization such as COPITEM. However, by the end of the project, COPITEM's membership had stabilized at 120 active members after successive periods of rapid growth and decline.

Geographically isolated, and in areas where political insecurity was even greater, the sheep farmers were difficult to reach. Project staff were challenged to overcome the farmers' natural reluctance to join groups. Of the three regional associations that were attempted, only one, AFORAM in the San Marcos area, was experiencing sustained growth at the project's end. A second association, CHIUL, had been abandoned due to local violence and project resource constraints. A third association, AFODAC, was not able to sustain membership, perhaps because farmers in this area were inexperienced in contrast to those of San Marcos who had had more contacts with other farmers in the outside regions. The lack of institutional development experience of FUNDAP personnel charged with the organization of the sheep farmers may also have contributed to the difficulties in this area.

Despite the best intentions and elaborate planning, the project suffered from limited co-ordination among project activities. There were few opportunities for regular communication among the separate field teams due to the disparate nature of the various activities for the three components, and the fact that teams were field based. Communication was further complicated by two factors: first, the lack of a full-time co-ordinator to provide a linkage among the teams, and second, FUNDAP's lack of experience in project implementation to fill that gap. The late 1980s was a time of rapid growth and change for FUNDAP. The organization was often stretched to provide the constant oversight needed to orchestrate the different interventions. The mid-term evaluation pointed to weak communication as a factor reducing the project's efficiency (Loarca, 1990:52). ATI, in its own trip reports, noted the need for FUNDAP to assume a stronger leadership role in overall project management.

A further problem was the difficulty encountered in changing the commercial relationship between the sheep farmers and the weavers. The project had envisaged the natural development of forward and backward market-driven linkages in the production chain. However, making improvements in the production of wool through improved breeding and flock care required considerably more time than initially assumed in the project design. Moreover, convincing the weavers to pay the higher prices for the improved wool was difficult. Furthermore, owing to deeply entrenched farmer habits and a lack of supervisory capability, efforts to improve traditional wool-washing practices had not been very successful and the grade of wool produced was not satisfactory. For all these reasons, the project had to be changed to include the construction of a wool-washing plant which would provide a direct link between the sheep raisers and the weavers. The plant, which was a significant addition to the cost of the project, eliminated intermediaries, thus reducing prices and further improving wool quality. Completion of the plant was hampered by construction delays and difficulties in forming a national sheep breeders' association to manage the facility. Construction of the plant was completed in 1993 and FUNDAP has taken on the management role. COPITEM has stated its willingness to co-operate in the operation of the wool-washing plant. ATI and FUNDAP have now prepared a plan to address wool washing, marketing and spinning aspects further.

A final point of relevance to this study was the project's lack of a gender perspective at the outset. The project had been started as a result of the request to FUNDAP by a group of weavers who, of course, were men. The project plans were based on the perception of sector linkages in which all the key end producers were male. FUNDAP and ATI were certainly aware that women were essential in the production process. For example, in the family weaving enterprises, women spin and reel the thread, help in preparation of materials prior to weaving, and do most of the finishing work on the woven product (washing and combing the nap). On the sheep farms, women help in the care of the sheep in pasture. The traditional family structure means, however, that the man makes the overall family decisions about production and marketing of the product, and receives whatever income results. The women were not paid any salary for their efforts and did not receive any share of the profits. Whereas, in Honduras, where the Cashew Project claimed to have been able to get women paid a wage for their contributions even though the men's groups controlled cashew nut production, and where a women's component, cashew fruit production, could be included almost from the start, in the Guatemalan wool project, this was seen as impossible. ATI and FUNDAP initially believed that they could not attempt project activities which disturbed the traditional family system, or their entire project might be compromised because male participants would resist and even withdraw from the programme. Efforts to get weavers to pay their wives for their work failed. In any case, the involvement of women was not a principal concern of the project.

Concern about the role of women, however, led ATI and FUNDAP to amend the project in 1991. This last phase of the project included additional funds and identified the need for new activities to promote the inclusion of women as direct beneficiaries. As a result, five women's groups were formed and over 90 female members of weavers' families were trained in income-generating activities of their choice. These activities, ranging from breadmaking and dressmaking to machine embroidery, began in mid-l991. The project hired a women's promoter in 1991. A separate study on the socioeconomic status of women in the sheep-producing areas was also undertaken in 1992 (Marina Delgado, 1992). ATI believed that the success of the project in mobilizing women in this last phase was due to the painstaking work with the men in the early phases through which enough confidence was developed to allow the revolutionary step of working directly with their wives.

Men and women react

Where, in the analysis of the Peru and Honduras projects, the direct developmental impact on women could be ascertained, here, because the project only worked with men in the wool sector, direct impacts on women can not be the focus of the study. Instead, the developmental impacts on men are explored and this analysis is combined with an attempt to ascertain the indirect effects on their wives of a project directed at men. In addition, a small sample of the women who were included in the project work in the latter phase are also studied to see what results their inclusion had for them. However, since their activities were disparate and not related to the overall focus of the project which was the wool subsector, they do not reflect the full thrust of the project effort. Questionnaires were administered, therefore, to thirty men from the project (shepherds, weavers), thirty men from the sheep and weaving areas but not associated with the project, and twenty-five female family members of project participants, divided among sheep and weaving project participants. Three women who participated in the women's programme also answered questionnaires.

Assessing indirect impacts is problematic (even assessing direct impacts is difficult!). It is hard, even if a control sample of women whose husbands were not involved is included, to be sure that the changes registered by the women are not due to other factors, such as efforts by Catholic action groups to mobilize women or the wave of changes spreading through the zone resulting from the reduction in violence and repression, and the improved economic situation throughout Guatemala. However, one advantage in this study was the involvement of Luz Marina Delgado who had extensive and in-depth experience in the zone, administered the surveys, and talked to the women at length about their experiences since the project had entered the lives of their families. A second advantage in this analysis helping offset the methodological problems was the mid-term evaluation done by M. Voorhips which was thorough, provided an additional source of information as to overall project impacts and is a reality check on the results which emerge here.

Our analysis above suggests that this project differs from those in Peru and Honduras in terms of the greater intensity of training in technology use, credit management, marketing and general organization as well in technology adaptation (and re-training) to ongoing enterprise concerns. There was also a very strong emphasis on forming producers' and marketing co-operatives to assist individual entrepreneurs in dealing with the market. The subsector approach adopted here requires intervention in all stages of production where possible efficiencies could be obtained. In other words, the project appears to have been more involved in working with the participants at all stages of their work, to solve their problems and to help them adjust to the actual economic circumstances in which they found themselves. Given this high level of continuous and intensive involvement, large impacts on the male participants might be expected and were, indeed, reported in the Voorhips mid-term evaluation. In his study, sheep farmers associated with the project proved to have sold significantly more than farmers not in the project over the same period, and earned a higher income as well. They were more likely to say their buying power had increased (Voorhips, n.d.).

In the UNIFEM survey conducted in 1993, the findings reported by Voorhips were clearly replicated. Participants queried had been positive about their experience in the various projects, but those in the Guatemala project were much more decisive than those in Peru or Honduras about the positive impacts the project had in all areas of their lives. Furthermore, the differences between the participants and the control sample were greater in terms of the relative economic advantages of the participants. Not only were these points true, but, in addition, wives of the participants reported substantial improvements for themselves, not only in income available but in changes in their role in the family and in their time use. This finding is all the more striking because reports indicated that women in the zone of the Guatemala project were working for their husbands without pay or other remuneration in the family enterprise, and had little status on their own. Thus, a men's project could have had the result of worsening the women's situation by increasing their subordination to the male head of household. Whether this is a result in this project is explored further below.

Looking at the specific findings, all of the men in the project and the men in the control sample came from rural areas and most fell in the 21 to 50-year old category. Most of the men had some education – more than three-quarters in the case of participants, more than two-thirds for the others. Of those who had had no education, sixty per cent of the participants had had literacy training while only eleven per cent of the control men had – a significant difference. More than half of both samples had wives who had no schooling, while the other's wives had limited primary school education. Three men in both groups said their wives had had literacy training.

About 70 per cent of the men in both groups said they were Catholic with the rest claiming other Christian adherence. A slightly larger group of the control men claimed to be without religion – 17 versus 10 per cent – but the number was in any case small. Ninety-seven per cent of the participants were married, as were ninety per cent of the other men. Fifty per cent or more of the men in both groups had four to six living children. Predictably, the two samples were about evenly split between farmers (sheep farmers) and those running a small enterprise (weavers). Equally predictably, both groups described their wives as housewives (27 per cent), farmers (28 per cent) or working in a small enterprise at home (45 per cent). Significantly more men from the project belonged to a co-operative; this difference, again, would be predictable as the project encouraged the formation of co-operatives and was especially successful with the weavers.

Forty per cent of the project men had joined six or more years before the survey, while another forty-three per cent had been involved for between two to five years. Fifty per cent said a new technology or the possibility of credit were reasons they became involved, and another third of the men said it was a combination of reasons including increased income and the provision of credit, a new technology, etc. Whatever their initial reason for joining, most men saw sharp positive economic impacts from the project. Sixty-eight per cent said they had changed their primary economic activity and increased their income (another eight per cent said they had changed their primary economic activity but that their income had not substantially changed). All who reported changed activities, including primary through tertiary activities, said the project had caused the change and had increased their income. Leaving aside the question of change, one hundred per cent of the participant men felt that the project had increased their income (68 per cent) or allowed them to get assets (25 per cent), or both (7 per cent).

In actual fact, there were economic differences between the participants and the other men. Both groups typically owned their house and land and were equally likely to have livestock – the sheep farmers did and the weavers did not. The participants, however, were significantly more likely to have other assets such as a savings account, machinery, etc – eighty per cent did, while none of the control group did. In addition, a higher percentage of the participants' wives had some assets either in the form of land, livestock or other assets. More than ninety per cent of those men in the project, who had assets other than house, land or livestock, had acquired them during the project and almost all believed the project had been the reason they had obtained the assets. A significantly larger number of men from the project were in the higher income category (based on a four-part ranking where US$800 to 700 is the middle income level, poor is next, and the bottom level is classed as below subsistence).

Like respondents in Peru and Honduras, the men from the project reported that their primary expense was on food, their secondary on clothing,

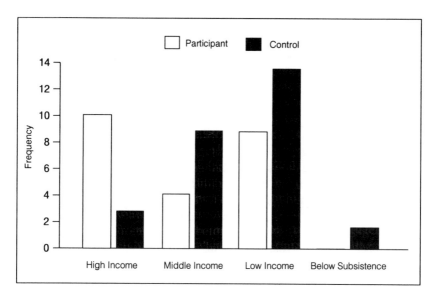

Figure 4.1 *Income Levels*

and education came third; this pattern had not really changed over the project period. Despite a very small difference in the two samples as to what they actually ate, the project men were convinced they ate better now as a result of the project because it had given them a higher income. They also said the project had allowed them to give their children more education than they would have otherwise, but this is not fully borne out by the survey results. Participants, consistently, over all their children, are slightly more likely to have more education provided, but the difference between the samples is not significant.

Consistently with the other findings, the men participating in the Guatemala project liked it. All but one said it had changed their lives. All said the project had trained them and ninety per cent of them said the project had provided either credit or a new technology or both. All believed it had either improved their income, or provided them with assets, or both. Seventy-three per cent said the family had benefited principally from having a larger income available for their needs, while another twenty per cent was divided between those who said the project had helped principally by allowing their children to get an education and those who said the main benefit of the project was that it had provided more work for the family. Virtually all the men said the project made them feel better about their economic future and about themselves. They also said their wives liked the project because (generally) of the money it provided. Few of the men thought there was a need to change the project in any substantial way.

Comparing the women who had been in the Honduras Cashew Project with the Guatemalan men from the wool subsector project, the overall picture is of a much clearer and stronger economic impact on the latter. Both groups said their income had improved but the Guatemalan men were *measurably* better off then the control men, to whom they were compared in terms of income and assets, while the Honduran women were not.

The three Guatemalan women who had been directly involved in the project and answered questionnaires had been in the project three years or less. All three were better off than the average since they reported having in their own name a house or land, livestock and other assets. Only one women claimed this relative wealth was due to the project. All said the project had led them to change their economic activities and increased their income and that this income improved their family's diet and/or children's educational opportunities. All said they had started to spend more time on work and less on their families and children. None of the women felt the project had had any impact on how the family made its decisions. Nor did they feel the project had affected their decision-making power, but they did think the project had given them a better outlook on themselves and had improved their economic situation.

The small sample and short time of project experience – and the distinctly different nature of the women's project from the men's – makes it difficult to interpret the female participants' responses as evidence of project impacts. These women seem to be better off than most women in Guatemala, in any case, and not due to the project necessarily. Nor do they report any other changes in their lives. Looking at the impact on the participant families in Guatemala through the eyes of the wives of participants gives a better picture, although this is, of course, reflected impact as the project has not worked principally with these women.

The wives are in the same age ranges as the men but they are substantially less well educated, seventy-eight per cent having had no schooling at all, and only four of these having had literacy training. All the women identify themselves as Catholic or other Christian denominations. More than ninety-two per cent see themselves as either farmers or as working in a small enterprise. More than half produce vegetables or poultry (eggs and meat) for family consumption and not for sale. The rest sell unprocessed products (like eggs and vegetables). None claim to receive any notable income from these activities. Only three of the women belong to a cooperative, although nine of them do belong to a women's group of some other kind. Half of the women say they help their husbands in their work; virtually none says she is paid for this effort. Most of the women (81 per cent) said they had helped their husbands before the project started just as they do now. Most of the women did not see that the project had changed the kind of work they did.

Most of the women, however, did believe the project had an impact on how they used their time but, in contrast to the project women in Peru and Honduras, these women were quite likely to say they now had *more* time for their families and their households. Half of the women said directly they now had more time to give to their families than before and that they now spent less time getting firewood or water. They said they had more time for rest and leisure activities, and three-quarters of them said they now had more time for their own education and development than they had had. All but three women said the project had changed their time use but, again, the majority who specified said it was because they now had more time for themselves and their family. This is interesting, and different from most of the women who had been in projects in the two other cases considered so far. Since these women still would like to have more time to do other things, principally their own work, it does not seem as if their perspective is completely different. Rather, providing more income to the household may actually have had the effect of lessening the existing burden on these women.

The wives report other favourable impacts on their families as well. Like their husbands, they report eating good food regularly and, like them, they say the project has provided them with a larger income so that they can afford better food. Their report of the family economic status – overall income and ownership by them and their husbands of all kinds of assets – is similar to the results reported by the male participants. Again, few women own anything – a striking contrast to the three project women! But the families are relatively well off compared to those of the male control group. Ninety-five per cent of the wives believe the project has affected their family positively. Most of them identify some combination of benefits for the family such as increased income or education for the children or a better life for the family. All state the project has provided a better income. Ninety-five per cent of the wives say the project has given them a good economic future, increased their own self-confidence or some other combination of positive impacts on their outlook.

What is perhaps most unexpected in these results is that many of the wives see their family status as having been positively affected by the project. More than half of those responding report that the project has given them more decision-making power. Specifically, about a quarter of the wives say they decide how to use their income while the rest make this decision in consultation with their husbands. For all of the other family decisions, the overwhelming majority of the wives say the decisions are made in consultation with their husbands. Asked if they have changed their decision-making role in regard to each of these matters individually, two-thirds of the wives say 'no', but a consistent third says 'yes', their husbands decided before and now they, the wives, are consulted. And, again, more than half the wives say the project has changed their decision-making role and given them more authority (while none of the three women who had been in the project say this).

In some ways, the responses reported here are contradictory. Eighty-one per cent of the women say they are still helping their husbands, just as they have always done, yet almost an equal number – 78 per cent – claim their decision-making power in the family has improved. Why should it have improved since the project provided them with no greater responsibilities than before? The answer may be found in the more detailed comments made by the women in answer to the open-ended questions.

First, although Luz Marina Delgado has underlined the disadvantages of women in this zone, the relative independence of Mayan women needs to be remembered. Unlike women in the other twenty-one indigenous groups of Guatemala, Mayans traditionally had many freedoms including the right to refuse a husband and rights to own land or go into business (Elmendorf, 1972). Over the years, under the influence of surrounding groups and the conservative Catholic church, many of these rights and privileges were eroded. Recent events, including the establishment of Liberation Theology groups, the civil war and migration from the poverty-stricken zone to find work (endured by both Mayan men and women) (Brandler, 1994) may have begun to reverse the downward slide in women's status. Mayan women in this zone are poor and disadvantaged but not totally submissive, both by tradition and because of recent experience. In addition, they are probably aware of outside events despite the lack of a pattern of women's group formation in this area. Closer looks at individual responses show some of this.[2]

For one thing, the wives wanted to have a chance to learn more in order to produce more and increase their earnings. One woman said, 'Even though this project did not give women direct participation in the project, it gave the opportunity for women to participate in other training. I would like to participate in productive projects with other women.' And another woman stated, 'We women have to learn more about techniques of soil preservation and to be promoters among other women.' Or, 'I have learned that women's organizations are important. We want projects to be directed to us. I would like support from a co-operative to take care of children so every woman can study without being distracted by them. Even if it were only two courses a year, I would like to be able to completely dedicate myself to study.' Finally, 'Now my husband gives me more money for food. This is good, but all they have taught women, like making soap and to embroider, does not produce. I want to learn productive skills.'

---

[2] Analysis of these individual responses was aided by Natalia Brandler. See Brandler, 'Indigenous Women and Strategies for Small Enterprise Development', unpub. mss. May, 1994.

Twelve of the women said they had learned new techniques for their own work through their husbands having been trained by the project, including sowing seeds, animal raising and food preparation. Now they had money to spend on better food for the family and on visits to their families. Both these things made them feel better about themselves. Two of the women said they now had time to take courses which they said made them feel good about themselves as well.

Many of the women felt the project had improved their relationships with their husbands. One said, 'The project changed my life. My husband used to drink but, with the orientation he received, he now brings the money to the house instead of drinking it.'

Other wives said:

'. . . now he recognizes and appreciates my work.'
'. . . he wants me to take courses and receive training.' (2 women)
'. . . he wants to know my opinions now.' (6 women)
'. . . he is on better terms with the family and with me.' (10 women)
'. . . he doesn't spend all the money drinking any more.' (4 women)

Many of the men in the project in fact recognized that they gained a new view of their wives as a result of their participation. Sixteen said the project had made them consider their wives as important and that they thought women should participate in courses and join co-operatives. They also said they have better communication with them and their family as a whole.

What we find in this study then, is that there is a striking and almost uniform pattern of the positive developmental impact of the Guatemala project on both men in the project and on women in their families. This does not mean that women have equalized their position. Clearly, in terms of their authority, education and economic opportunity, they have not. But they have been mobilized by their experience – even though only through their husbands – by the project's training, outlook and economic impacts. Thus, not only is their income improved along with a lightening of their domestic workloads, but they have raised their sights for themselves. Furthermore, they have improved their family position somewhat – many men now take them less for granted or even encourage their further education. The project has also had an impact on one of the severe problems faced by women in this zone – drunkenness and the ensuing violence of their spouse. Perhaps as a result of feeling better about themselves and their own economic future, the husbands are no longer drinking away what few family funds exist. Family communication is better. Women can only benefit as a result.

The Guatemala project may be faulted for not having included women in its first phases, but it did have positive impacts on the wives and children of those who did participate – a 'trickle-down' effect in economic well-being, in lightening workloads, in status, and in the mobilization of expectations.

This can not be taken as evidence that projects working directly with women are not important or necessary. Women's work skills and options have to be increased or their overall development will never take place. The wives interviewed here express this strongly; they needed and wanted to be trained and supported in their own work. It is worth noting, however, that this successful subsector project, with its intensive interventions at all stages of the economic process, had a positive impact in virtually all dimensions of the lives of the men *and* women in the households which became involved.

The bottom line

The Guatemalan Wool Production and Processing Project is still completing some final work and, indeed, the scheduled official completion date was only three years ago (May 1992). The project was a relatively large, complex and expensive undertaking for an NGO in a very difficult area. Yet, despite the short time since completion, the relatively short project period and the complexity of activities, the project has registered significant successes already. In all three component areas the primary objectives of the project were achieved, although FUNDAP and ATI believe the work with the weavers to have been the most successful aspect. The major achievements in each area are presented below:

Sheep and wool production

Prior to this project, one of the key problems facing weavers in their processing had been the low quantity and poor quality of scoured wool. To confront these problems, the project worked with sheep farmers to upgrade their breeding and wool production methods. The project interventions not only improved the raw material needed by weavers but the overall productivity of the sheep farmers. Measures of project achievements include an increase in wool output per animal from 1.5 pounds a year to 4 pounds a year with improved sheep stock, an increase of about 18.5 per cent in the birth weight of lambs and a decrease in the mortality of lambs from 35 per cent to 3 per cent. As indicated above, this meant a substantial increase in farmer income.

Wool processing and weaving

The principal activity of this component was the enhanced production of quality wool goods through such technical and business improvements as expanded market opportunities, the acquisition of better materials, improvements in production, the establishment of quality and pricing standards, and the development of a sustainable market. The institutional formation of a viable weavers' co-operative was a significant achievement according to both ATI and FUNDAP. The rise in participating weavers' incomes and the stability of the new weavers' co-operative are major indicators of this component's success.

Commercialization
The project established INNOVA (Innovations in Handicrafts), a marketing enterprise specializing in the commercialization of Guatemalan crafts. It also assists weavers and other artisans in the development and marketing of new product lines. Among its achievements were the establishment of a market distribution system, entry into national and international crafts and tourist shows, the conducting of market surveys, establishment of a local retail outlet and the cultivation of relations with other handicraft institutions in the country such as museums and the Guatemalan Tourist Institute. INNOVA adopted a quality control system and gave marketing assistance to COPITEM for its direct sales. INNOVA twice won the award for best exporter of Guatemalan handicrafts. In the first three years of its operation, it reached annual sales of $90 000 from a first year figure of $4000.

Outside evaluators have noted and praised the project's successes. The mid-term evaluation pointed out that the project had achieved what it set out to do, and more, in all three areas (Loarca, 1990), but it also did suggest some areas of concern, such as that the relationship between the farmers and the weavers had still not been worked out. Weavers had not yet accepted the necessity of paying higher prices in order to get a higher quality. Thus, farmers were still not experiencing an increased demand for their higher quality wool. This failure to increase the demand for better quality wool may be the reason the project has not been able to get broad representation among sheep farmers, or a successful sheep farmers' association. Farmers are still inclined to rely on traditional practices and relationships, including those with middlemen.

Additional personnel problems, mentioned above, continued to plague the project. It would not be reasonable to fault FUNDAP, with its limited experience in projects of this type, for not having prepared a network. Yet in conceptualizing the project, the human resource demands of such extensive and complicated activities were perhaps not taken into consideration as much as they should have been.

The final and most important area of criticism for this book is the initial lack of attention to the role of women in the subsector, and the impact which this project might have on them. Both ATI and FUNDAP noted that women were not included at the outset but explained it on the grounds that, in such a conservative and traditional environment, any efforts to work with women would be abortive and hurt other aspects of the project as well. They argued that they would have been unsuccessful in attempting to work with women had they not spent the first phase of the project building trust and confidence among the men. This claim can not be tested one way or another. In Honduras, in a traditional area, it was possible to work with women from the outset. But given the peculiar conditions of the Quetzaltenango region where suspicion and distrust of outsiders was so high, it may have been more difficult. In any case, women were ignored at the outset and, by giving them no role, their subordination in the production process was reinforced.

Later, in the final phase of the project, women's activities were initiated; 90 women were drawn into five groups for a variety of activities. However, an additional question may be raised about what was then done for women when this element was finally addressed by the project. It is interesting that, with a women's project co-ordinator finally hired, the five women's groups finally established, and with a women's promoter in place, they worked on whatever the women in the group wished to produce. On the one hand, this is being responsive to the wishes of the participants, but it is a rather sharp contrast to the method of approach adopted by FUNDAP and ATI in the men's project. In the latter case, an activity which was shown, through feasibility and marketing studies, to be economically viable and in the long run self-sustaining and replicable was supported. The women's programme seems to be much more casual, thus exposing the women to some of the same problems noted in Peru, including the potential lack of a reliable market for the products the women chose to produce. However, such problems have not surfaced yet and a recent ATI evaluation suggests that the women's programme has been able to successfully incorporate women into the local economy, a substantial departure from the norm for this zone.[3]

If the Guatemala project is judged in terms of the goals the collaborating agencies set for it at the outset, it had a remarkable set of successes after just a very short time. Its record supports the subsector approach and the broad systematic assault on structural impediments up and down the whole chain of activities which produces the final market product. This project had a major impact on ATI's approach to development work. It will have an impact on work done by other agencies as well, where often limited intervention in one microenterprise leads to frustration and inefficiency as the real block to improved production lies somewhere else down the chain. While implementation of a full subsector approach may require greater financial and technical resources and a longer period of time than a narrower project concerned with a single intervention, the prospects for significant and sustainable economic benefits are much greater. It may be an essential approach to resolving the full problems of production and marketing in many product areas and thus providing the most adequate support to small entrepreneurs.

---

[3] Comment by Carlos Lola, reported to Lucy Creevey by Eric Hyman in comments, April 1994.

An additional positive example exhibited by this project is the successful collaboration and overall monitoring which took place. In the study of Peru, we found that certain local agencies did not have the experience, the resources, or the orientation to effectively support the women in the food processing project. In contrast, ATI and FUNDAP worked very well together. Both shared the same philosophy and approach to development.

The bottom line is that women benefited from the Guatemala wool subsector project but, not having had their needs and interests centrally considered, the benefits may have been more limited for these women than had they been the focus of project activities at the outset. The inclusion of women would, of course, have increased the complexity and cost of an already expensive project which was difficult to manage and co-ordinate as it stood. However, women are often key links in the chain of production and, in future projects, application of the subsector approach should – if improving the position of women is a goal – require directly including women despite cultural barriers and increased cost.

Chapter Four: Statistically Significant Survey Analysis Results
Table: Guatemala

1) Sample type (participant, control) and literacy training (some, none)
   Chi sq. = 4.866, Phi = .506, P. = .0274
2) Sample type (participant, control) and membership in a co-operative (yes, no)
   Chi sq. = 29.792, Phi = .78, P. = .0001
3) Sample type (participant, control) and ownership of other assets (yes, no)
   Chi. sq. = 38.212, Phi = .812, P. = .0001
4) Sample type (participant, control) and level of family income (high, middle, low)
   Chi sq = 7.232, Cramer's V = .377, P. = .0269

CHAPTER FIVE
Bangladesh – Surjosnato Coconut Products

Poverty and hope

Bangladesh is one of the poorest countries in the world with a population of 106.7 million in a land area of 144 000 km². Population density is more than 740 people per square kilometre. This situation is not likely to improve soon because, despite extensive family planning programmes, only fifteen per cent of the women use contraception and the total fertility rate in 1990 was still 4.6. Average per capita annual income is $210[1], much lower, of course, in the rural areas. Life expectancy is only 52 years. Only 16 per cent of the population live in cities, the rest are rural dwellers. Although agriculture provides only 38 per cent of the gross domestic product, a majority of those economically active get their livelihood from this sector. Because of population density and a heavily skewed ownership of land, most rural dwellers are landless and must sell their labour to survive; jobs, however, are scarce. Ninety-four per cent of the rural population lives below the poverty line. Bangladesh is particularly subject to flooding which regularly creates major suffering, and loss of life and property in its rural areas (World Bank, 1992:216,218,222,270,278; Heyzer, 1988:1110).

Bangladesh, formerly East Pakistan, broke away from Pakistan in 1971 and established a democratic republic based on 'four pillars': democracy, secularism, socialism and nationalism (Jahan, 1991:197). However, quarrels among the élites and the inability of the government to improve the severe economic situation led to a military coup in 1975, after which the emphasis on secularism was reduced. Indeed, the military regime had to form an alliance with the radical Muslim fundamentalists, including those in the Jamaat party, which led to an amendment in 1977 substituting for the pillar of secularism 'an absolute trust in Allah' (Jahan, 1991:239).

The low rate of economic growth relative to the growth of the population (the GNP grew 0.7 per cent between 1965 and 1990) (World Bank, 1992: 218) has left Bangladesh heavily dependent on international loans and donor support. In 1990, the total external debt was 54 per cent of the GNP (ibid:264). As a result, many international agencies and donor countries have programmes in virtually every sector of the Bangladeshi economy.

---

[1] The national currency is the taka. The exchange rate used here is 38.1 taka = $1.00.

One area in which the government has allowed and supported extensive efforts is the informal sector. Rather than concentrating efforts through a ministry of its own, the government has permitted action to go principally through private operations and a large number of NGOs with informal sector programmes have proliferated in the country. The most notable of these is the Bangladesh Rural Advancement Committee (BRAC). BRAC, an indigenous organization, began as a small relief organization in 1972 following the war of independence. It is now the largest NGO in Bangladesh and one of the largest in the world with a paid staff of 2500 people, with headquarters in Dhaka, the capital, and it has seventy centres spread throughout the country. Its funds are all from external donors, the major ones being Canadian, Norwegian, Swedish, Swiss and German aid agencies, with smaller amounts from UNICEF, NOVIB, NORAD, Ford and other foundations. Its operating budget in 1986 was approximately $40 million (Lovell, 1989:134).

BRAC's primary focus is to empower poor men and women by mobilizing them to work together in co-operative groups. This means training not only in the use of a technology or in business skills, but also in group management and co-operation. It is a multi-faceted approach requiring more involvement than a purely business approach would imply. BRAC has two major programmes. One of these is a Rural Development programme (RDP) in which there are four primary subprogrammes, each with a slightly different mix of activities. All, however, follow the same basic purpose which is the organization of landless men and women into co-operative groups which 'plan, manage, and control collective activities and facilitate individual endeavours.' The objective is to bring about social and economic improvements in their lives and make them as self-reliant as possible. The normal strategy is to form male and female groups in each village, help them with functional literacy programmes and consciousness-raising activities, provide them with training in such things as leadership, management and income-generating skills, and provide or help them to acquire other tools needed to make them self-sufficient. By the end of 1986, BRAC had reached 114 000 villagers: 60 000 women and 54 000 men. Through its credit support of the RDP, between 1979 and 1986, BRAC had loaned out about $37 million (Lovell, 1989:135).[2]

BRAC appears to have far more impact on changing the lives of the rural poor than does the national government, which BRAC views as 'weak and inefficient'. Indeed, BRAC has gone beyond its original mission of working with the poor to trying to change the national government setting by conducting training courses for government officials to sensitize them to the problems and potentialities of the disadvantaged population.

---

[2] BRAC also has a major initiative in primary and preventative health-care activities nationwide and has assisted in establishing non-formal primary education centres which, by 1987, had reached 30 000 children, the majority of whom were girls.

Despite a scatter-gun approach which demands a large amount of money and trained personnel, it has been remarkably successful. Evaluations of BRAC, although identifying some problems in the execution of its programmes, generally concur that it is effective and has had a major positive impact (Lovell, 1989:131–156; Chen, 1987; Beets, Neggers and Wils, 1987:15–20). BRAC is an interesting counterfoil to the Coconut Project examined here.

The growing forces of conservatism: women in Bangladesh

The situation for women in Bangladesh is a very complicated and difficult one. One primary factor is the growing influence of the radical Muslim militants who, through their religious ideology, attempt to constrain women from public roles (83 per cent of the population is Muslim, and 16 per cent is Hindu) (USIA, 1992). Purdah, or the actual seclusion of women, is still practised in many parts of Bangladesh. In Ramganj, the site of the Coconut Project, purdah is kept very strictly. Interestingly, only wealthy families can afford to fully practise purdah, keeping the wives from being occupied with any but domestic tasks or in employment opportunities that only put them in contact with females. Being a teacher or a doctor to women students or patients is, for example, acceptable under purdah. At the level of the poor rural women, families rely on females' economically productive work, but restrictions are still imposed. Women are not allowed in public spaces such as fields, markets, roads and towns, but must remain 'secluded in the private sphere – homestead and village, from which they emerge only at prescribed times and for prescribed purposes' (Chen, 1987:74).

The quotation from Marty Chen below gives a good description of the situation and work of poor rural Bangladesh women:

> Like women elsewhere, village women in Bangladesh work long and strenuous days. They raise and tend animals; thresh, parboil, dry, winnow and husk the grain; grow fruits and vegetables; clean and maintain the huts and homestead; give birth to and raise the children; and, occasionally, produce crafts for sale or home use. However, unlike women in other areas of intensive rice cultivation who are actively involved in transplanting, weeding and harvesting, village women's work in Bangladesh is confined exclusively to the post-harvest activities of threshing, winnowing, drying, husking, milling, and storing grains. Therefore, the only wage labour traditionally available to women in rural Bangladesh is post-harvest or domestic work in other households. Moreover, unlike other countries where women play very important roles in trade, rural women in Bangladesh seldom leave their villages for the markets, either to buy or sell. As a result, the few Bangladesh women who engage in trade do so only at the lowest levels – as petty hawkers in their own villages. (Chen, 1987:74)

Women in Bangladesh have begun to challenge the traditional patterns but resistance to change is very strong. Even the ending of purdah is not universally supported by women. Because purdah has been fully possible only for the rich, it has acquired a value for women with few other opportunities for status in their society. The reluctance to give this up still persists among some (Jahan, 1991:240–241). Women are beginning to break down the restrictions, but slowly. Education is one of the chief reasons for this change. As of 1990, 78 per cent of all women aged 15 to 24 years were still illiterate, as were 87 per cent of the oldest age group (UNIFEM, 1991:52). However, 64 per cent of appropriately aged schoolgirls were attending primary school; eleven per cent were in secondary school and only four per cent were in institutions of higher learning (World Bank, 1992:274). Education is gradually becoming accessible to a wider group of women.

Women have begun to move out of their private home space into paid jobs but at a very low rate. Only thirteen per cent of Bangladeshi women were reported as economically active in 1988 (although this low figure must be questioned because of the universal tendency to disregard women's productive activities as being 'economically productive'). Seventy per cent of these were in agriculture, ten per cent were in services and twelve per cent in industry. Most women in industry were in the export-processing zone enterprises working on line jobs for low pay and no benefits. Twenty-eight per cent of the industrial workforce are women (Heyzer, 1988:1110, 1117).

There is a women's movement in Bangladesh; at least, there are numerous women's groups organized to fight different aspects of the situation facing women in the country, but they do not operate with a unified agenda. There are some prevalent themes, such as ending violence against women (beating and rape) which is widespread in society, but they are not otherwise unified in their concerns nor in the strategies they propose. Nonetheless, the continuing growth in the number of these groups and their increasing public exposure is a good sign for the future of women in the country (Jahan, 1991:240–244). Equally important, and certainly closely related, is the success of BRAC and other NGOs in helping to set up women's co-operative groups with multiple economic and social functions. There is still stiff opposition within society (which may be why BRAC is seen as less socially motivating than its sister organizations in India and Sri Lanka, AWARE and the Sarvodaya Shramadana Movement, as it has an even harder social climate to breach) (Beets, Neggers, Wils, 1987–8). A large number of women, however, have been touched by these efforts and the word is spreading.

The Surjosnato Coconut Project

The Coconut Project was begun in June 1980 with ten producers. At this time, a Mennonite Central Committee (MCC) programme officer and a Canadian food technologist were working in the Maizdi region (see Chapter Five, Annex for a description of MCC on page 109). They found that coconuts could be easily processed for drying in an area where the nuts were very abundant. The ensuing project was developed after a year of experimentation and training at the Maizdi centre. The purpose of the project, which was established under the MCC's Jobs Creation programme, is to 'create employment in Bangladesh primarily among poor rural women'. In this area, many, if not most, men are unemployed and landless, so providing jobs for women could have a major impact on their status as they might become the only wage earners in the household. Even if they are not the sole wage earner, it is unusual enough for a woman to attain an income-producing activity in this impoverished area; what the women involved in such a project might, as a result, be expected to experience is a serious alteration in their lives. The principles on which the project was based include:

1. Promoting and assisting businesses which:
   a. are operated and managed to benefit the producers;
   b. are replicable, viable and sustainable;
   c. have the potential to become financially and legally independent and self reliant;
   d. are based on employment-expanding products and processes; and
   e. have diversified product lines.
2. Providing training to project personnel chosen because they:
   a. have no or a low income,
   b. are landless with few or no assets, and
   c. are primarily rural . . . (MCC, 1992:1–2)

The first ten women, selected at the suggestion of leaders in three local villages, were widows or single with no male earning an income in the family, and were poor, owning no land (except their homestead). In 1981, the project opened its office at Ramganj and increased the number to 22 participants from six villages. They recruited the project manager at this time, Mir Hossain, a man who had done mechanical and electrical engineering for the army and lived in Ramganj. Three other men worked on the project staff at this time.

The tasks of the project were several. The first of these was to teach the women how to produce dried coconut using a home solar drier. Solar driers had existed in the area before the project, but they were used only for home consumption. Dried coconut powder was purchased from Sri Lanka or the Philippines, or coconut oil was used. The women had to be trained to construct and use an improved solar drier and to process the coconut to a marketable quality. Women visited the project centre every ten days or so to obtain and husk an average of 300 coconuts. They then took the coconuts home where the flesh was removed, grated, sulphured and dried. During the dry season, they processed around 30 coconuts a day.

Marketing was done by the centre. First, however, project staff had to create a market. The demand for dried coconut was already there, but imported produce or local fresh coconut satisfied that demand. MCC targeted some of the larger businesses, such as candy and biscuit makers, to get them to purchase the project's dried coconut. Another MCC project, Source, was the marketing outlet and helped with the marketing strategy, including door-to-door selling, distribution of free samples and price dumping. MCC had to build a road for access to the project and this was accomplished in 1982 with the assistance of a food for work programme.[3] The women were also given the materials and helped to build their home solar driers. The project took repayment from the first production cycle. The women did not really understand the process of applying for, being given, and repaying a loan, and did not know how to buy the materials for a drier on their own.

A second activity was introduced in 1983 because of marketing problems with the dried coconut product. Women were taught to produce coir from coconut fibre which could then be used for such things as ropes, brooms, brushes, nets, wall hangings and carpets. The coconut husk had to be crushed by a mechanical, electrically powered crusher so that the fibre for these coir products could be extracted. The women hired for this aspect of the project worked at the project centre for the duration of their work.

The women who produce coir are paid a wage for their work. The women who produce dried coconuts get dividends after the costs of the project (initial purchase of the coconuts, paying the project staff, maintenance of the centre) are met. Gross income to the project is divided: 10 per cent for the managers; 50 per cent for dividends, debt reservicing and the centre costs; and 40 per cent for the salaries of those working on crushing and coir products. Currently, a typical income for a woman working at home drying coconuts may be as much as $18 a month, while a coir producer might earn about $15 (average income is less). Designers constitute a third type of employment, and their salaries differ from the other two groups. The project became self-supporting in 1984, although MCC continued to provide supervision and technical support.

One problem that led to a major change in the dried coconut aspect of the project was the humidity in the rainy season which prevented drying. This resulted in a serious loss of income for the participant women. It also threatened their market, for the project could not provide a steady or

---

[3] Food for work programmes have been mentioned in reference to all the Latin American countries studied here as well. Under such a programme, beneficiaries contribute their work (in this case building a road) in return for food. This approach has been criticized by all the international agencies we are studying here, as well as by local agencies with an emphasis on a business approach or on self-reliance as a goal, such as FUNDAP and MCC, because it creates a dependency mentality.

guaranteed amount of product in this season (coir production could go on all year as it is done inside at the centre). As a result, MCC technologists went to England and consulted with ITDG. This, in turn, led to the contribution by the latter of an electrically powered tray drier (the pump and fan are electrically powered, heated by wood), costing approximately $860 to construct. The drier was installed and in operation by 1987; and that rainy season saw a significant increase in the production of dried coconut (ITDG, 1988).

The introduction of the drier highlights an interesting characteristic: the reliance of one aspect of the project on the other. The drier itself is relatively expensive to operate because of its need for electricity. Moreover, revenues produced by the women who process dried coconut are not sufficient to cover the cost of operating the drier. Combined with the revenues from the coir, however, the returns are adequate. In effect, the coir production is subsidizing the coconut production. As a result of the introduction of this method of augmenting dried coconut producers sales, coir producers receive lower salaries (ITDG, 1988). However, more women are engaged in dried coconut production which has the attraction of allowing them to work at home most of the time. Thus, by tying the two activities together in this way, the Surjosnato project is economically sound and able to employ more women than if it consisted of only one activity.

The initial contribution from MCC was $26 247 in addition to an interest-free loan of $1680. In the succeeding five years, MCC made cash contributions of *about* $20 000 a year and further larger interest-free loans (in 1985 they loaned $132 940 to the project. After 1985, MCC ceased cash contributions although it continued its technical support and backup.[4] This project was totally managed by MCC. ITDG did not have oversight of this project but it did contribute the tray drier and provided technical consiultation to MCC technicians.

Dealing with the impossible

The Surjosnato Coconut Project has met and overcome many difficulties along the way. The first of these was the severe poverty of the zone and the lack of resources to support project efforts. Even things as straightforward as transport of both the raw materials and the finished product to market remained an obstacle, despite the construction of a road in the early stage

---

[4] Because we do not have exact figures showing how much MCC contributed to Surjosnato or what its staff costs were, our statements here are estimates. The figures we cite are too low and do not even include ITDG's costs or specific contribution. Better figures are needed for a closer comparison with other projects, however, it is clear from the information we do have that this was a very low-cost endeavour relative to the Peru and Guatemala projects. Like Honduras, but even more so, this was a long-term project which required small contributions from the sponsoring agency.

of the project. The price of raw materials was a further problem as the supply of coconuts varied depending on the amount of rainfall. In 1992, for example, the number of coconuts available during the peak season fell drastically and the price rose accordingly. The project had no means of substituting on any large scale and suffered accordingly.

A third problem which hampered the project, particularly at the outset, was the strict traditional attitudes of people in the zone, particularly the Muslims. Initially, Muslim men were opposed to allowing their wives to participate in the project. The project, in part because of this constraint, chose older, widowed women at the outset but later began to draw in married women as some of the resistance lessened.

A fourth and major problem was, and remained, finding a reliable market for the dried coconut. The largest competitor for the dried coconut product is fresh coconut which is preferred by most consumers in Bangladesh and available much of the year (although there is a definite coconut season). A foreign market was never really established. In fact, Sri Lankan and Philippine dried-coconut products are cheaper and, some think, of better quality than those produced at Surjosnato, even though, after having quality problems in the beginning, the project has managed to get the women to produce acceptable, high quality dried coconut. In 1983, the project faced such financial difficulties that it almost closed down (losing $1890 in that year). Instead, it reduced the price of its coconut product and cut back on the number of coconut producers who could participate to the 20 who appeared to produce a product of better quality and better sanitation than the others. Out of the remaining 38 producers, 15 were chosen to be trained for coir production.

However, the market for new coir production was also uncertain. The original designs of products were not popular and the machinery used was inadequate. The project had to hire two coir specialists (a supervisor and trainer) and a coir machine processing operator. Gradually the production capacity and the quality did increase. In addition, the defibring and by-products enabled diversification so that this aspect of the project could cover its own costs and even raise the salaries for the coir workers. Coir processing began to subsidize some of the costs of the coconut drying. (It is estimated that 20 to 40 per cent of those costs are paid by the coir production.) The coir market itself was not stable and the project faced competition from local homemade products. Prices had to be kept low and the market did not expand rapidly. In the two years before 1993 it was not possible to increase the salaries of any of the workers in the coconut processing or coir sections. In response to the marketing and raw material supply problems, the project manager, as discussed above, decided to diversify activities and the project began to grow coconut plants, vegetables, etc.

Although the coir production seems to balance the coconut losses at the present time, other innate problems which prevent the project from establishing its overall goal of self-reliance for the women producers remain. The first of these is the dependency of the women involved in the project on the project itself and its staff. Women coconut producers are technically shareholders just as the coir producers are employees. Both are represented on the Management Committee of the project, but most of the management staff are men and all operations are directed and controlled by the director. There is a women supervisor, but she is more involved in quality control than in management. It would be unusual for the female participants in the project, many of them Muslim and brought up in an extremely patriarchal environment, to argue with the male staff. Furthermore, the women (who have had only some primary schooling, if any education at all) do not have the skills necessary to read and understand financial documents. Thus, they do not understand the management of the enterprise and consider themselves employees. They do not even know how much they received in dividends or savings the previous year. Empowerment may have resulted from the project at the home or village level but, contrary to the intentions of MCC, the enterprise does not belong to the women even though they are shareholders in it.

Nor is the project staff itself independent from MCC although Surjosnato is running with its own manager and paid staff and is supposed to be an independent enterprise. That manager, however, still appears quite dependent on the representative from MCC. He (the manager) understands well the functioning of the operation and is able to monitor the transactions throughout the year, but he still (as of 1993) relies on the MCC representative even to the extent of needing him to draw up the balance sheets. Furthermore, Surjosnato remains dependent on the Source for its marketing. Sixty per cent of its produce goes through this outlet. This makes project independence highly questionable.

A final problem which the project has had to endure is the national unrest and political turmoil of 1992-3 which led some foreign NGOs to cease operations in the country. The Surjosnato project was apparently not directly affected in its home operation, but efforts to expand the market or to maintain the sales levels may have been undermined.

Because market problems persisted, the project introduced a coconut nursery to sell plants, and began growing vegetables and other plants for sale as well. Other possible side activities include poultry raising. However, the centre still holds its own in terms of income and MCC has replicated the project in Jobarpar (Barisal district), where 23 women have been involved since 1991 making coir products, ropes and processing coconuts. In 1992, there were 106 women involved in the Surjosnato project: 60 drying coconuts, three working in the coconut section, 21 defibring, 19 making coir products and three on the management staff. Ten men were engaged in the

Table 5.1: Surjosnato activity results to 1992

Description	1986–7	1987–8	1988–9	1989–90	1990-1	1991–2
Total workers:						
coconut workers	50	48	48	60	60	58
coir workers	28	29	33	44	51	49
supporting staff	9	9	9	9	9	10
Average earning per month: (taka)						
coconut workers	537	546	665	665	674	674
coir workers	275	323	361	401	335	396
supporting staff	1 067	1 318	1 563	1 945	2 202	2 279
Rent	24 800	25 700	28 200	42 100	42 300	47 280
Maintenance	4 084	11 762	16 958	6 893	15 988	10 000
Sales:						
coconut (taka)	2 974 028	2 759 0371	3 727 0478	4 501 0113	3 644 0637	3 471 2233
coir (taka)	125 807	195 544	228 716	399 454	312 101	316 955
other	57 985	33 466	46 121	153 679	173 731	305 266
Profits:						
(taka)	119 103	135 795	137 087	63 380	102 022	69 875
Profits distribution	59 551	67 897	68 543	31 690	51 011	none
Cumulative capital	85 608	221 246	446 631	733 642	809 868	939 606
Loan balance (MCC)	400 655	325 000	275 000	200 000	180 000	109 000

project working on coir. There was a staff of nine to run the centre, provide extension and quality control, handle marketing and the crushing process and tray drier, and so on.

Women speak out

To indicate, as does the discussion above, that the Surjosnato project is on fragile ground is not to say that it has had no impact on the women associated with it over the last thirteen years. This project took place in a zone where not only women can not find jobs, but men as well are landless and often unemployed. The association of the women for a period of up to thirteen years with a programme which has enabled them to receive a regular income might be expected to have had some impacts, not only in terms of the revenue they received, but in regard to other aspects of their lives.

The sample studied through the 1993 survey included thirty-three women who worked with the project, both coir workers and coconut home processors, and a control sample of thirty-two women. Our analysis indicates that there were numerous significant differences between the control group and the participants. In the first place, most of the women in the project came from rural areas, while more of the non-project women came

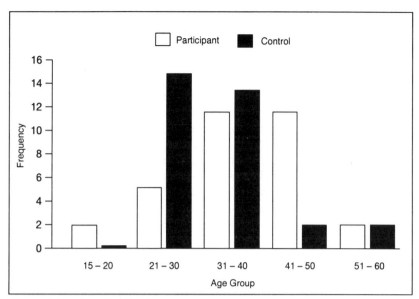

Figure 5.1 *Age distribution*

from villages. Both groups were in close-by areas and from a section of the population seen to be more or less economically equivalent, but the participant women were older than the control sample, a significantly larger percentage of them falling in the forty-one to fifty year old category (consistent with the project having chosen older women, at least at the outset).

More than two-thirds of both samples indicated they were not heads of their household. Most women in both groups were married. More women in the project than in the control sample had four children but the difference between the groups in regard to number of children was not significant. What was significantly different, however, was religious affiliation. Perhaps not surprisingly given the restrictions on Muslim women, the participant women as compared to the group of non-involved women were more likely to be Hindu than Muslim, three-quarters of them being so. The sample of women not in the project was also more likely to be Hindu than Muslim, but a larger proportion of them were Muslim. The disproportionate number of Hindu participants indicates the difficulty experienced at the outset by the project in getting Muslim families to allow their women to work.

A plurality of the women in both groups had never been to school but almost forty per cent of both groups had attended some or all of primary school and fifteen per cent had had more education. Some of the participant women, however, had had literacy training (20 per cent) which again significantly distinguished between them and the uninvolved women, none of whom had had this experience.

One of the more striking and significant differences between the participants and the other women was their declared occupation. All of the project women identified themselves as being primarily employed either in a small enterprise or (for three women) as a teacher. Eighty-one per cent of the other women said they were housewives, and their primary occupation was care of their homes and families. This distinction certainly reflects the general lack of economic opportunities for women in this region. In contrast, most of both groups of women reported their husbands had jobs, although the two groups varied in what jobs their spouses held. Most of the women who were not in the project said their husbands worked as paid labourers in agriculture, while most of the husbands of the project women worked in a small enterprise either at home or outside the home for a wage. A greater – and significant – difference appeared between the levels of education of the husbands in the two groups. Forty per cent of the non-project sample had had no education while this was true for only fifteen per cent of the participants' husbands.

Most the women had joined the project more than six years before. All of them said they joined because it offered them the possibility of earning a higher income. Their hopes had apparently been borne out. Eighty-seven per cent of the women said the project had either given them a higher income (50 per cent), allowed them to acquire assets which they otherwise would not have (22 per cent), or both (16 per cent). Eighty-six per cent said that they had changed their economic activities because of the project and were now earning a higher income. When the two groups are compared

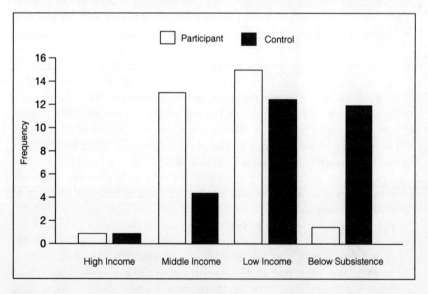

Figure 5.2 *Family income*

objectively in terms of current economic status, the project women are clearly better off. In regard to family income, significantly more of those working in the project fell in the middle-income category. Fifty per cent of the participants are considered low income, and only six per cent fall below the poverty line. Women in the control group, in contrast, fall forty per cent below the poverty line with another forty in the low-income category.[5]

There is a difference between the groups in regard to family assets although not quite as striking or clear cut. Most of the husbands of both groups own land or a house, few have livestock and well over ninety per cent of the men have no other assets. Most men do not get an income from their assets, but the women not in the project are significantly more likely to report that their husbands get at least some income from their assets. Most women in both groups do not own land or a house (84 per cent of both groups), but non-project women are significantly more likely to own livestock (consistent with they and their husbands being more likely to work in agriculture). Participant women, however, are significantly more likely to have jewellery, the traditional form of savings for women – more than ninety-seven per cent do, while ninety-four per cent of the control women do not.

Whatever the difference in income or assets, both groups of women say that their primary expenditure is on food and that that has been their primary concern since before the project started. Nonetheless, most (70 per cent) of the women in the project believe that the project had an impact on their expense patterns because it provided them with more income.

In regard to other measures of family well-being, it appears there is little actual difference in diet, but all the participants still claim the project has improved their family diet and that, as a result of it, they eat more and better food now than they did. More than 80 per cent of them also claim that the project has resulted in their children getting a better education, but this is not fully borne out by a comparison with the children of the control women. It is the case that a larger percentage of the non-project children (controlling for rank order of children) had had no schooling and, for second, third and fourth children, a larger *participant* percentage also had had some secondary education, but the differences were not statistically significant.

In terms of the overall impact of the project on how they lived their lives, women in Bangladesh were somewhat more ambiguous in their responses than the family members in Guatemala. These women said that they spent the same time on care of their household and their families as they always had (88 per cent). They also spent the same time getting water or gathering fuel, although most of them said that they had less time to rest now or for

---

[5] The scale here is based on an annual income measure with taka 5000 to 7000 equaling low income.

their own leisure (79 per cent said this, with the rest saying there was no change). *All* of the women said they spent the same amount of time as ever on their own self-development or education, which is certainly a clear reflection of the lack of emphasis in this project on mobilizing or training women other than in the specific job required by their work. Indeed, none of the participants belonged to a women's group or to a co-operative, while the non-project women all belonged to a co-operative and a few also belonged to a women's economic group.

Most of the participants felt the project had affected their use of time but they were not specific as to how except for a third of the women who explained that they now spent more time working. Interestingly, the participants did not want to change their use of time while most of the other women who answered the question did (they wanted more time for work).

In regard to decision-making, the comparison between the two groups is curious. Project women were more likely to say they decided on the use of their income, on what they worked and how they worked, on what the family ate and on what education their children – both boys and girls – should have. In contrast, the other women were more likely to say their husbands made these decisions, although they were likely to at least participate in decision-making in regard to family diet and the education of girls. In addition, they were far more likely to say that the husband alone decided on where they lived as well, although many participants also said their husbands decided this.[6] Nor did these differences coincide with religious differences. Muslim women in the project were also more likely than their non-project counterparts to have more decision-making power. Most of the participants (53 per cent) said the project had given them more power than they had had in family decision-making. This is striking because these same women also said in regard to each specific type of decision that there was no concrete change over the project time. This implies the women could not pinpoint exactly what had changed, but that there was change in the direction of their gaining authority. This is sharply reflected in the greater authority they appeared to have when actually compared to the non-project women.

All of the participant women were enthusiastic about the project and its contribution to their lives. All felt it had improved their outlook on the future, and most of them said this was because it had given them more confidence in themselves. Eighty-seven per cent, as stated above, said the project had given them a higher income, allowed them to acquire assets, or both. All but one of the respondents felt the project had had a profound

---

[6] The overall pattern of difference between the respondent, the husband, some other family member or the respondent and the husband making decisions is only significant at 25 per cent across all decisions, which is not statistically strongly meaningful, yet it is clearly the control women who are more liable to say their husbands make the decisions.

positive impact on their family. Most (79 per cent) identified this contribution as providing the family with more income, but a distinct group (18 per cent) said the project's major impact on the family was their own increased independence. Asked directly whether the project had an influence on their family authority, eighty-two per cent said that it had given them more decision-making power and all (100 per cent) said the project had given them a more positive outlook either on themselves or their economic future or both. All of the women stated their husband liked the project, most for the income the family now received. Asked directly, a sample of husbands of the participants corroborated this, eighty-one per cent saying they liked the income which was now available to the family. The few who answered they did not like the project did not specify what they disliked, although the marked change in family authority may have threatened some.

Interestingly, most of the project women had no suggestions for changing the project, 66 per cent saying it was fine as it was. A few asked for more or longer support from the planning agency, others having a variety of suggestions with no clear pattern. For most women, it was clear they had no criticisms, only an appreciation of what the project was doing for them and for their families.

Comments by individual women show something of what the project had meant to them. It should be stressed that these comments came from both Muslim and Hindu women and from those working at home and those producing coir products at the centre. Such feelings appeared to be universal across the participants. One woman said the project had let her save money for her marriage. Another said that before she had spent her time in idle gossip but now she is able to earn an income instead. She also said that before she started work, her family ate once a day, two or three days a week, and now they could eat three meals a day and even have fish three or four days a week. Another said the project gave her social importance and many women said that now 'they had money to spend as they wished'. One woman said that the work was too hard and made her ill, but she said she would continue because she needed the money. She said the work had helped her to provide schooling for her daughters and that her husband now sought her opinion more often. Several others also said that they could provide necessities that the family could not have had before and several specified the education of their children. One participant said her earnings had kept her family from starvation and another that she no longer had to beg for credit.

In sum, the Surjosnato project, which has provided its participants with up to thirteen years of experience, has had a major impact on the women who participated in it. These women may not have had their conscience raised in the sense that they were mobilized to recognize their rights and their potential to achieve such things as self-reliance in their work or equality of status in society. Their isolation and the lack of emphasis on mobilizing them as a group is reflected in some of the responses here. More striking by far, however, is the progress which this group has made. In a region where there are no economic opportunities for most women and where even the men are faced at best with un- or under-employment and marginal salaries at best, providing the women with a steady income has had a major impact on their families and on them as well. In the first place, they (and their families) are better off economically than a random sample of non-participant women. Since these women were not recruited for ability or former association with this type of enterprise, but rather were initially chosen on the basis of need (see project description above), the advance of this group to being in the low to middle-income category is indeed an achievement. This achievement now places the family well-being, if not mainly, at least substantially in the income-earning capacity of the women. As a result, the women appear to have taken on a larger role in family decisions than women generally do. They may work harder than they did before, but their position is acknowledged as more important. Moreover, Valerie Autissier points out, based on interviews, that men were less likely to beat their wives now that they earned an income through this project. This is an elaboration of the meaning of these survey results. In terms of long-term impacts, then, although it has not mobilized the women to take charge of their own work and enterprise, Surjosnato has had a major positive developmental influence on those women involved in it. This impact may be far greater than other projects will be able to achieve in countries where women generally have more opportunities and where the economic climate has more openings for men as well, not excepting those projects with more training and mobilization for the women built in. In Ramganj, this was a revolutionary departure and it had a major and profound result.

Balancing the results

The major accomplishment of this project must be the creation of 106 jobs in an area where jobs are extremely scarce and the economy very difficult. This enterprise has been able to adjust to market fluctuations and to continue to pay off its loans and other expenses, and enable an ever increasing number of rural women to earn an income which would be more than they could have otherwise earned (in some cases they would have been unable to earn anything at all). Moreover, its methods bear a similarity to that which has been successfully adopted by BRAC (discussed above). The poorest women are chosen, those with no regular income, no land (except perhaps their homes), and no skills. These are not successful small entrepreneurs looking to expand or improve their activities; these are people who have not been able to find an economic niche for themselves and are on the margin of survival. They are uneducated and unskilled. These women are drawn into an enterprise which is tailored for their rural lives. Many of them do most of their work at home, which is customary. The coir workers, who work at the project site, are primarily women and are often either very young and single or older than the average (and are more likely to be widows or single heads of household), so their work outside the home is also acceptable in society. Like BRAC, then, the Surjosnato project is reaching women among the poorest of the poor and finding ways they can be employed in their own economically productive activities. Moreover, it is moving on to use one project as a model for others.

The costs of this work appear relatively low, although there is a large hidden subsidy in the form of continuing MCC work, at least in part based on volunteers. In the final evaluation, however, three major problems remain, at least one of which may eventually defeat the project. In the first place, the market for dried coconut products is very difficult. There is great competition both from local fresh coconut and from imported dry products. Surjosnato has managed to find a market (largely through the MCC marketing outlet, Source) but that market is not expanding very rapidly. Indeed, Source itself saw a net loss in 1992 of 126 104 taka ($3310) (MCC, 1992). Surjosnato has been unable to get into the export market in any serious way. In the past, this was because the quality of the product was not high enough although that problem has been rectified. Also, Sri Lanka and other exporters who process coconut on a much larger scale are still able to sell their quality product at a lower price. ITDG in 1988 estimated that the Surjosnato dried coconut was three times the world market price (ITDG, 1988). This limited market outlook, combined with the fluctuating cost of the raw material (coconuts) and rising costs of operation, make this project hardly an advisable choice for many other groups of women as it may not be sustainable in the long run. ITDG notes, in fact, that 'the prospects for expansion appear limited' (ITDG, 1988). Even the coir production does not obviate this conclusion. Coir products too have had a limited market and, of course, coir production employs fewer employees than coconut production. The expansion of the whole operation in response to some major market shift does not appear likely. The economic future of Surjosnato, then, is shaky although the enterprise is diversifying activities and attempting to adapt to the market changes and economic climate of each year.

The second major problem is the dependency of the women discussed above. Women are still dependent workers, which is inconsistent with the goal, identified at the project's inception, of making women the owners and managers of their own enterprise. The project staff talk about further training and workshops to give them the skills to understand the business thoroughly and to enable them to take a leadership role. If some kind of intensive socialization training to enable the women to take a position of real active membership and even direction within the enterprise was launched, this could possibly transform the situation (although not the market!). Such training is difficult in the social context of the area, but not impossible since the women are working in the project (and as BRAC has managed to do in some of its projects) (Chen, 1987:73–97). This is a serious flaw in the project, although it could be rectified in theory. Actual transformation into a women-controlled venture, however, is unlikely. The women themselves have not articulated any desire to increase their role; this would be outside their customary role pattern. Furthermore, thirteen years of experience as employees, with little responsibility beyond their own specific production, does not prepare them to take charge easily.

Finally, the project, after thirteen years, is still quite dependent on marketing, the technical backstopping and even the oversight provided by its progenitor, MCC. MCC plans to make the Source an independent enterprise. This raises the risk that the marketing outlet will not find it profitable to continue to sell coconut products which have not been as successful as export sales of furniture and other designed products. It is not clear whether the Surjosnato project could survive on its own without such help.

Yet the Surjosnato project has had a major impact on the lives of the women who have worked in it. These impacts are substantial, not only in the area of income and assets, but also in terms of improved family status and feelings of self-worth. A final evaluation of this project must balance the improbability of substantial sustainable replication in a wider community in Bangladesh, and the uncertain future for these women in the project now, with the real gains they have enjoyed in the past.

Annex – Chapter Five

The Mennonite Central Committee (MCC) was founded in 1920 as the 'relief and development arm' of the Mennonite and Brethren in Christ Churches of Canada and the United States. Currently over 1000 volunteers from Asia, North America and Europe are involved in programmes covering areas such as agriculture, community development, job creation, emergency relief, education and health in over fifty countries. In August 1993, the MCC elected a woman to head its entire operation, the first time that a woman would hold that position.

The MCC first came to Bangladesh in 1970 following the great tidal wave disaster of that year. Later, programmes other than relief were introduced, and there are now three foci to MCC work in Bangladesh: agriculture and family development, employment creation and emergency assistance. Self-reliance is one of the basic principles of MCC projects, so things are not given free. Instead, MCC works with rural poor people, especially women, to create jobs through which they can support themselves over the long run, and also trains the staff of existing development institutions. It hires and trains national staff for its projects and uses the work of international volunteers to support the process. MCC has its own direct projects and works with other NGOs and national agencies through co-operative liaisons. It also differs from many NGOs in its emphasis on establishing viable economic enterprises and its stress on marketing. MCC objectives for 1992 were to increase the number of jobs in its direct and co-operative liaison projects by ten per cent. Surjosnato and the recently established Jobarpar are not the only direct MCC job creation projects. The others include a handicrafts enterprise at Saidpur, a screen printers' enterprise, a letterpress division, a handmade paper project, a palm handicrafts project, Badga enterprises making woodcrafts, rope and twine, and a marketing centre (the Source). The enterprises to which MCC gave liaison support in 1991-2 included carpet making, food processing, pottery and doll production, dried fish production, potato chip production, brass products, mat making, and clay pot and lamp production. MCC Job Creation has supported at least seven other similar programmes. The Job Creation Programme has a staff of 31 people, including eight in their Appropriate Technology division who are, among other things, food technologists, appropriate technology specialists and engineers. Other members of this programme staff include, aside from the overall programme administrator and assistant, training specialists, design technicians, administrators and budget and financial specialists, export-marketing specialists and office managers (MCC, 1992).

Chapter Five: Statistically Significant Survey Analysis Results in Bangladesh[7]

1) Sample Types (participant, control) and dwelling (rural, village)
   Chi sq. = 10.036, Phi = .393, P. = .0015
2) Sample Type (participant, control) and age (15–20,21–30,31–40,41–50,51–60)
   Chi sq. = 14.171, Cramer's V = .467, P. = .0068
3) Sample Type (participant, control) and religion (Muslim, Hindu)
   Chi sq. = 3.015, Phi = .219, P. = .0825
4) Sample Type (participant, control) and literacy training (none, some)
   Chi sq. = 3.672, Phi = .369, P. = .0553
5) Sample Type (participant, control) and occupation (housewife, runs or works in small enterprise, teacher, public sector employee, retired or unemployed, other)
   Chi sq = 61.128, Cramer's V = .97, P = .0001
6) Sample Type (participant, control) and husband's education (none, part primary, all primary, part secondary, all secondary)
   Chi sq = .11.534, Cramer's V = .506, P. = .0212
7) Sample Type (participant, control) and membership in a co-operative (does not belong, belongs)
   Chi sq. = 65, Phi = 1, P. = .0001
8) Sample Type (participant, control) and membership in women's economic group (no, yes)
   Chi sq. = 3.351, Phi = .229, P. = .0672
9) Sample Type (participant, control) and family income (high, middle, low, below subsistence)
   Chi sq. = 11.217, Cramer's V = .425, P. = .0106
10) Sample Type (participant, control) and husband's ownership of other assets (none, few, many)
    Chi sq. = 5.368, Cramer's V = .358, P. = .0683
11) Sample Type (participant, control) and respondent's ownership of livestock (none, some)
    Chi sq. = 3.983, Phi = .26, P. = .046
12) Sample Type (participant, control) and respondent's ownership of other assets (none, some)
    Chi sq. = 53.597, Phi = .908, P. = .0001
13) Sample Type (respondent, control) and respondent's receipt of income from assets (none, small, substantial)
    Chi sq. = 8.197, Cramer's V = .376, P. = .0166

---

[7] Although this annex reports only statistically significant results, the decision-making comparisons are also reported here because of the interpretation of the emerging pattern in the text.

CHAPTER SIX
India – The Sericulture Project

India and development

India is a vast country with a very poor population, and its 849.5 million people live in a land area of 3 288 000 km² with an average per capita annual income of $350 (World Bank, 1992:218). The Indian economy has been growing and expanding; GDP, for example, went up 3.6 per cent between 1965 and 1980, and 5.3 per cent between 1980 and 1990. The structure of the economy has also been changing from a primarily agricultural base to one increasingly dependent on industry and the service sector. In the early 1980s, 44 per cent of the GDP was produced by agriculture; by 1990, 31 per cent of the GDP was from agriculture, 39 per cent from industry and 40 per cent from the service sector (World Bank, 1992:220, 222; Gopinath and Kalro, 1985:284). The majority of the population is still engaged in agricultural activities, however, and poverty is widespread. As recently as 1985, it was estimated that 48 per cent of the rural population lived below the poverty line. Inflation is a factor in India although it is not as detrimental as in the Latin American cases studied here. From 1980 to 1990, the inflation rate was 7.3 per cent (Heyzer, 1988:1110).

The Government of India, assisted by many external donors, has made serious efforts to control the growth of its population. Significant progress has been made – as many as 45 per cent of the women in India use birth control measures and the rate of population growth has decreased from 6.2 per cent per annum in 1965 to 4 per cent in 1990. However, the numbers of people added to the population each year is still very large; indeed, the population of India has increased from 361 million in 1951 to 843 million in 1991. Thus, although the percentage of people living below the poverty line has declined from 55 per cent in 1960–61, to 30 per cent in 1990–91, the number of people living below the poverty line increased from 167 million in 1960–61 to 211 million in 1990–91 (World Bank, 1992:270; UNIFEM, 1992:1).

The area of the project discussed here, the Udaipur District of Rajasthan in the northwestern part of India, is one of the poorest in the country. This is where the Bhil tribe, the target of the project, live. These dry foothills of the Aravalli range are barren and droughts are frequent. Whereas previously forest products supplemented the agricultural produce, they are now rapidly disappearing. As a consequence, rainfall is decreasing and arable soil is eroding. The agricultural yield is very low and there are no large industries or other establishments to provide employment. In order to survive, the people have had to get credit through money lenders. Often, they have to provide free labour to those middlemen because they have no means of repayment. Some Bhil have migrated to try to find work in the cities, but the others remain, increasingly impoverished, dependent on occasional projects, such as government road construction where they can sometimes work at a minimal wage.

India is a democratic republic governed by a parliamentary regime. Since independence, in 1947, it has followed a path of centrally planned development with the government continually making efforts to support the weaker sectors of the population and provide incentives to all sectors of the economy. Although the principal focus has been firmly on the formal sector and the growth of industry, the government has recognized the importance of agriculture and, more recently, the informal sector in which such a large percentage of the population is involved. In India, some 90 per cent of the working population is 'self-employed', that is, either home-based workers producing garments, textiles and other products, or small petty traders and vendors, or the providers of services, and those who provide manual labour in agriculture, construction, catering, laundry, etc. In this group, at least 60 per cent are women. Indeed, Indian government statistics show that 94 per cent of the working female population is self-employed (Bhatt, 1988:1060–61).

The Indian government assists this sector of the population through various programmes in different ministries and agencies, and through the work of numerous NGOs financed by external donors now active throughout the country. In 1987, the Prime Minister of India established a National Commission on Self Employed Women and Women in the Informal Sector, whose task was to study the socioeconomic status of women in these areas of work and recommend policies to improve their lives and work. In 1988, the Commission submitted its report which, among other things, called upon the government to set targets for women's development at the local level and recommended that they actively initiate and maintain a network of grassroots-level organizations such as village-level women's groups (Bhatt, 1988:1064–5). This foundation of government awareness of the informal sector, and particularly the needs of women, created a favourable environment within which the most recent phase of the Sericulture Project developed.

Indian women

The Government of India has shown increasing concern for the plight of women over recent years. As early as 1971, a National Committee on the

Status of Women was established to consider all questions pertaining to the rights and status of women. In 1977–8, the Planning Commission established a Working Group on the Employment of Women. Working Groups on the Development of Village Level Organizations of Rural Women and on the Role and Participation of Women in Agriculture were also set up. Other commissions and groups considered adult education for women, women's studies and women's access to science and technology. And, as mentioned above, in 1987–8, a commission considered the position of self-employed women as well. In the seven national plans adopted by the government, attitudes toward assisting women gradually changed. Some major governmental decisions attempted to equalize the place of women; from the outset they were granted equal political and civil rights. Outstanding atrocities such as *sati*, the practice of burning the wife at the death of her husband, were outlawed. In the first plans, however, women were primarily assisted through welfare measures with a gradual and ever-stronger emphasis on access to education for women. By the Seventh Plan, although programmes including both formal and informal education continued to be important, there was more focus on empowering women to use the skills they had and to acquire more so that they could be successful in their economic endeavours. This included training women to make them more competitive in getting jobs in non-traditional fields. It included policy initiatives requiring equal pay, exploring part-time work opportunities and prohibiting some of the worst continuing features of societal discrimination against women, such as the dowry system in which men had to pay the families of the girls they married and then extracted work from their spouse to pay off the debts they had thus acquired. One of the major concerns in the Seventh Plan closely relevant to this project was the emphasis on village-level organizations for women, called Mahila Mandals.

These government policies and programmes had a positive impact on the situation of women in India, a situation which had been an extremely difficult one. Throughout history, women had not been treated as equal parts of society. Governed by men in both Hindu and Muslim societies, women had few rights or privileges and many obligations. Married at a very early age (twelve or younger in the countryside), they are expected to do all the domestic work and contribute to the economic production of the household as well. For most women, owning land or other property is impossible as these are controlled by their husband. Indeed, women have such an inferior value to the family that girl babies are often unwelcome. As a result, one of the ironical outcomes of the campaign to reduce fertility is the unbalancing of the sex ratio through the growth of abortion clinics in large towns and cities where women go for amniocentesis and, if the examination indicates a girl, an abortion. Between 1978 and 1982, 78 000 female foetuses were aborted (Desai and Patel, 1985:15–24). Women are also often subject to violence. Observers estimate that as many as two million women are raped each year, most aged 16 to 25 (approximately 78 per cent of the victims were unmarried). Wife beating is also common, seldom reported, and virtually never punished by law (Ghandi and Shah, 1992:36–51).

Education, the prime factor in changing societal attitudes, is gradually spreading throughout India. Eighty-two per cent of primary school and thirty-one per cent of secondary school-age girls are now enrolled in school (World Bank, 1992:274).[1] Where 81 per cent of all women over 25 are illiterate, younger women are substantially less likely to be without schooling (60 per cent are illiterate) (UNIFEM, 1991:52). Rural areas remain far behind urban areas in this regard although, among the Bhil in Rajasthan, women are less likely to be illiterate than the national average.

Women are heavily disadvantaged in their economic roles throughout India. Although women in poor families are universally occupied with more than domestic work in an effort to feed, clothe and shelter their families, few find paid employment or are registered as 'economically active'. Where 52 per cent of the men are wage earners, only 20 per cent of the women are. Of these, 81 per cent are involved in agriculture, 11 per cent in industry and 9 per cent in services. Again, women are commonly in the worst-paid and lowest status jobs.

As elsewhere in Asia, most of the poorest people are in the rural areas, and women form a substantial proportion of the poorest of the poor, especially women whose husbands have died or migrated and deserted them (Heyzer, 1988:1110–11). Nor has the major reform of Indian agriculture, the Green Revolution (which, in the 1960s and 1970s introduced a new, higher yielding strain of rice, better farm technology, irrigation and chemical fertilizers) particularly benefited women. There is more call for women's labour as increasing yields demand more weeding, yet this is a rather negative development. Women are most in demand in areas where landowners have, because of the increased rice productivity, evicted their tenants and resumed farming themselves. Where the demand for women workers has increased, their wages have not. 'In fact, the household incomes of landless agricultural labourers are lower where there are higher concentrations of women workers because women are normally paid much less than men'. Women work longer hours in the fields and have higher workloads than men but only get 40 to 60 per cent of male wages. They are given the most labour-intensive tasks such as weeding, transplanting and harvesting (Heyzer, 1988:1112). In the Udaipur district, of course, there has been no increased demand for women's agricultural work since agricultural production has been in decline.

There are numerous women's organizations in India which support efforts to improve the position of women. One of the major organizations

---

[1] The figure is for 1989. Five per cent of young women enter higher education.

is the National Federation of Indian Women (NFIW) which has organized rallies against price increases (1981) and pressured the government for action against the dowry system and violence against women. The NFIW has offered literacy programmes to rural women and has led other campaigns that support rural and urban women in many different ways. Other organizations have sprung up out of the specific activities of women's groups. One of the more successful of these is the Self Employed Women's Association (SEWA), organizing 40 000 self-employed women. Their programmes (with external funding) have included loans to poor self-employed women. SEWA was the chief promoter of the government's commission on self-employment discussed above. Other unions and professional associations for women have developed, especially in the late seventies and eighties, through which women in the member groups have been able to exert pressure for changes in their job conditions, wages and benefits, etc. (Desai, 1988:38–41, 84–87, 193–202; Ghandi and Shah, 1992:276–84). However, membership in such movements is commonly made up of the élite and educated women, or women in specific wage categories. Poor rural women (such as the Bhil women) are much less likely to have been touched by their campaigns (Sharma, 1991:159–210).

The Women's Sericulture Project

The Government of Rajasthan, through its Tribal Area Development Department (TADD), established the Integrated Development of Women in Sericulture Project as one response to the abject poverty in the Udaipur district. This project was part of an overall national programme entitled 'Increasing Employment Opportunities for Tribal Women'. The Udaipur Sericulture Project aimed to generate additional direct employment for some 300 tribal women in twelve villages in two blocks of Jhadol and Girwa through a variety of activities including mulberry tree cultivation, silkworm rearing, silk reeling and the production of appropriate equipment for the enterprise.

The choice of sericulture was based on two principles. First, the market for silk seemed inexhaustible, both inside and outside the country. Second, viable alternatives to generate income in the zone were difficult to find. Sericulture is an agro-based and labour-intensive industry which is suitable for the traditional pattern of landholding and work in the zone. It requires only a small investment to start. Although sericulture was not common in Rajasthan because of the wide variation in climate and scant rainfall, it was possible, especially in the project belt where rainfall is slightly higher than elsewhere. Studies by the Agricultural College at the University of Udaipur indicated that this enterprise was viable, and a local NGO, Rajasthan Vidapeeth, experimented initially in a few villages to prove that the undertaking could work.

In 1983, a three-year project was implemented by TADD with the assistance of UNIFEM. Sericulture was a new activity in the region and all TADD staff in charge of monitoring the project had to be trained in Mysore. The project began slowly in the Jhadol and Girwa blocks, taking 25 beneficiaries per village and adding two new villages per block per year. All beneficiaries received one month of technical training in sericulture. The number of beneficiaries reached 300 by the end of the first project phase in 1985. All were chosen with a minimal requirement that they (their family) had at least 0.1 hectare of land and access to water.

UNIFEM's contributions covered the costs of mulberry cultivation, silkworm rearing, equipment, vehicles, compensation for crop losses and a revolving fund for purchasing cocoons from the producers. Rajasthan government inputs were primarily for infrastructure and administration costs, such as buildings and salaries.

Initially, Bhil resistance to the project was strong. It was difficult to convince the people that they should not pursue their traditional cropping patterns but should, instead, use 0.1 hectares of their land for mulberry trees, especially since these did not produce for at least one year. They feared the loss of their subsistence without being confident of the benefits which they would eventually see. To overcome this barrier, the TADD staff tried a variety of approaches. Chief among these was a contribution of 400 rupees (US$15) to cover the cost of the crop foregone by the mulberry planting. TADD also used traditional opinion leaders such as village chiefs to persuade the others, and even sought to convince the doubtful with oaths by project officials before a religious shrine in a Pai village (Bakht, n.d.). These methods and the possibility of increased revenue overcame most opposition and, increasingly, mulberry trees were planted throughout the project area. Both men and women in the family worked on the planting and care of the mulberry trees, but women solely were supposed to be kept occupied with the silkworm rearing and silk reeling. During the fourteen days when the eggs are hatched and silkworms emerge, the eggs have to be at a constant temperature and humidity in a special communal rearing hut. The larvae are later cared for in the house itself and require round-the-clock feeding until they spin a cocoon around themselves. These rearing activities, oriented as they are to the home, are compatible with the traditional domestic duties of the Bhil women.

Marketing of the final cocoons was done entirely through government institutions supervised by a committee composed of the Director of Horticulture, the TADD programme officer, a member of the State Administration and the Deputy of Horticulture at the state level. This committee decided the purchasing rate for the cocoons and then auctioned the cocoons obtained.

TADD reactivated a women's co-operative, the Women's Co-operative Society, which had existed to produce local cigarettes (*bidi*) and provide tailoring services, but which had gone bankrupt. Members of this co-operative were taught how to make bamboo equipment for the silk enterprise, including the little trays in which the worms are raised in the home. The co-operative was given a start-up grant for its activities and it prospered. There are 153 women currently involved in this activity.

In 1987, UNIFEM conducted an evaluation of the sericulture project. The study found the sericulture project an economic success. Families could earn from $25 to $33 per crop of worms. From a 0.1 hectare holding, a farmer could earn $100 – three times what he would have earned from his traditional crops. The women who began reeling silk thread could make substantially more. 'Many families earned as much as Rs10 000 (US$310) annually.' (Bakht, n.d.:7). Other aspects of the project were less healthy. Economic growth had been achieved without necessarily improving the lives of the people, or so the report claimed. In particular, the Bhil had substituted mulberry trees for their traditional crops. Now they faced shortages in their normal foods. Ill-balanced diets and malnourishment were common, leading to further ill health. Women who participated in the project did not seem to have gained in self-esteem, nor did their children benefit from the possible advantages of the project such as better food or increased opportunities for education. All too often money was being spent on what the report called 'luxury items' such as household utensils and silver jewellery. Alcoholism increased dramatically for both men and women (Bakht, n.d.:8).

In addition, although this was clearly intended to be a women's project, the report indicated that women were not fully enjoying the benefits. Men, who owned the land, were integrally involved in raising the mulberry trees. It was men who attended the meetings held by the TADD project officers and men who went to the community rearing huts at night to feed the silkworm larvae. Men continued their traditional role as heads of household in directing the use of new monies which became available, whereas the women beneficiaries of the project, unused to organizing themselves as a group to defend their own interests, continued as they had before in a position of total subordination to their spouses (Bakht, n.d.:9–10).

Because of the evaluation's conclusions that many of the developmental impacts were not desirable, UNIFEM did not participate in the project for four years (1986–90). In this period, the Government of Rajasthan continued to carry the project without support, extending progressively the number of villages where the Sericulture Project was implemented until as many as 3000 beneficiaries were reached.

In 1990, a new phase of the project was begun, again with UNIFEM's support. Two hundred new beneficiaries were targeted, bringing the total participants to 500. The focus of activity, however, was not to be on the individual women but on the women's groups, Mahila Mandals, so that the women could be mobilized to understand their own self-worth and their own potential for action, both as individuals and as a group. These women's groups would be trained and organized in such a way that they were to become the focal point for developmental activities in the villages:

> In addition to sericulture, these groups (would) plan, implement, and monitor supportive income-earning activities, the health and nutrition of the beneficiaries and their children, encourage school-going habits amongst the children, inculcate the habit of savings amongst the participants and generate a joint fund in cash or kind to cater to the consumption needs of the group. (UNIFEM, n.d., 2)

As the project became more diversified and mushroomed in the region, UNIFEM established linkages with various organizations. The project was still implemented under the control of TADD, but UNIFEM obtained the collaboration of the Department of Women, Children and Nutrition (UNFPA), the World Food Program (WFP) and a non-government organization, ASTHA, which was already working with women in tribal groups in the zone. TADD undertook the management and administration of the project and paid for administrative costs and technical training. UNIFEM provided funds for purchasing inputs and supplies, such as saplings, and for blasting wells and other activities. ASTHA developed awareness-raising activities for the target women who were organized in Mahila Mandals. ASTHA's activities were financed by UNIFEM as well. The WFP provided 1635 days of 'food for work' to initiate mulberry cultivation and the building of rearing huts and community centres for the Mahila Mandals. Out of the Rs22 spent per worker per day, Rs5 were paid in cash to a worker, Rs11 were given to buy raw material, and Rs6 were given back to WFP to buy 2kg wheat, 200g of pulses and 200g of oil to be given to each worker each day.

As technical training, assistance, purchasing and marketing by TADD continued, ASTHA implemented the organization and training of six Mahila Mandals. Training in awareness and leadership was organized, as well as community activities. Poultry farming was encouraged as a training experience. Mushroom growing was introduced and so was the cultivation of vegetables in association with the mulberry trees. The latter was particularly successful as the vegetables not only protected the soil from erosion but could be sold locally and added to the nutritious food supplies available in the zone. The Mahila Mandals also received 'kisan nurseries' composed of bamboo and mulberry saplings from the Forest Department to add to their income. Mulberries provided fodder for the silkworms and firewood while bamboo was used to make equipment for silk-worm rearing and other utensils, furniture, etc.

By 1992, the project had five hundred participants but many more families in the zone had been drawn into the broader sericulture project. The costs of the project remained relatively low ($523 994) given the large number of people reached and the 'spread factor', that is, the way this enterprise has expanded beyond the actual project participants through TADD's continuing work in the zone (see Annex 3 for details on project costs).

Establishing the reality

The Sericulture Project met and overcame numerous problems. The first of these was the resistance of the Bhil people to the new project, and this was handled largely through subsidizing the first year of participation (see above). The silkworm enterprise proved economically viable and participants' family income showed a significant increase. The average family budget went from $32 a year to $226; given no other alternatives in the area for income generation, this increase remains extremely attractive. In other parts of Rajasthan, however, where other opportunities exist, sericulture has not taken hold because the potential earnings are not in fact that great, especially given the large amount of labour which this enterprise requires.

Care of mulberry trees, silkworm rearing, and reeling silk require a very simple technology, the use of which is easy to teach even to illiterate people, although substantial knowledge and care (which cannot be taught so easily or quickly) is required to keep trees and silkworms free of disease. The market for the products which the Bhil women produce does not seem saturated. Furthermore, the participants are able to respond to demands for increased quality, and cocoons of both A and B level are being marketed. Thus, from this perspective, the long-run future of the project seems good. Somewhat in conflict with this conclusion is the possible result of an opening of the Indian economy to the world market, as urged by the World Bank and other donors. At present, TADD purchases the cocoons and silk thread and sells them, providing the participants a guaranteed price for their products. If the Indian market is opened, the price may be driven down, which, given the size of the profit margin for the individual families, may threaten long-run viability. In fact, Ila Varma reports that in May 1993 when the silk yarn from TADD was put up for auction, there was no buyer at all because the price was higher than that of cheaper and finer-quality silk yarn now readily available from China. In addition, the World Bank is sponsoring a sericulture project in seventeen states of India which emphasizes a higher, more sophisticated, technology. While UNIFEM and the World Bank have engaged in consultations about their different perspectives on a sericulture programme which may influence the larger World Bank programme and be of use to the TADD growers and reelers, there is still a potential range of competitors who may be able to produce more silk thread, of higher quality, with less labour and eventually undermine the Rajasthan operation.

Aside from the economic future of the project is the central question of the changing role of women. The work of ASTHA seems to have made a difference to the women participants organized in the six Mahila Mandals. Reports indicate that women from these groups now go to meetings to learn how they can do things for themselves and are assisted through *balwadi* or creche programmes for their infants and toddlers and *anganwadi* programmes for pre-schoolers. They have had training in health and nutrition as well. ASTHA has used plays and puppetry to teach its lessons and took group members on field trips to see what other groups of women had been able to achieve. Eventually the experience of the women in the Mahila Mandals has borne fruit. Four of these groups became very active. They report reprimanding teachers who did not show up, questioning the provision of inadequate food to children in the *anganwadi*, and demanding the food and produce they had requested from the local ration shop. They challenged a corrupt village head and refused to vote for him, and discussed the practice of wife swapping and how it could be terminated. In other words, they were beginning to take charge of their own affairs and were becoming a political force at the local level (Bakht, n.d.:19). Certain groups also began to turn aspects of the silk business into small group enterprises, such as a commercial worm-rearing centre and a silk-reeling plant (Bakht, n.d.:19–20).

In addition, the women's co-operative successfully produces and sells bamboo products and has begun to grow vegetables. Not all the women's programmes worked out as intended, however. A poultry project failed because the poultry houses were too far from the women's homes and the breed of chicken chosen was not well adapted to the zone. Mushroom cultivation was also unsuccessful because of the lack of a ready market for the products. Furthermore, most of the women, even in the Mahila Mandals supported by ASTHA, have not asserted themselves strongly or changed their normal activities or their relationships with their dominant male family members. Some progress, however, has occurred in the four mobilized groups mentioned above.

A final by-product of this project was the establishment of close collaborative relations among a variety of different agencies with different agendas which were all now sharing a common focus on the women's sericulture project. This established (or at least reinforced) a precedent of sharing information and resources among disparate groups in order to achieve a common end. Not surprisingly, of course, this collaboration was not always easy. ASTHA staff point out in interviews that they sometimes had problems with TADD. Some of the male fieldworkers, they state, were well-intentioned and interested in helping the Bhil women, but others were insensitive to women's problems, even trying to make money for themselves from what the women were doing. It would have been better to use female extension workers, they state. ICDS (the children's service programme) conflicted with TADD over certain issues, such as the placement of a village centre and the use of space in another centre by the Mahila Mandals which had been supposedly allotted to the ICDS *anganwadi* programme. In another instance, the WFP found itself without a room which TADD had promised it (Bakht, n.d.:23). TADD, on the other hand, criticizes ASTHA because it reaches so few women and does not appear to be able to help the large number that need their programme. All these conflicts were mediated, and all parties have been able to work with each other productively, but certain inconsistencies in outlook, organization and patterns of implementation remain.

Impacts on women

Questionnaires were administered to forty-three women who had participated in the project, twenty-five from among the original participants and fifteen who had joined at a later point. Twenty-eight women who were not involved responded to questionnaires. In field notes, the enumerator, Ila Varma, reports on many of the women participants who answered the questionnaires. Some, she says, do not seem to have a very active part in what is being done or to have changed their lives much; their husbands clearly still make all of the decisions. For a majority, however, she observed changes in their households resulting from the project. Growing new crops (like cabbage) is one example and, in another, the women built the worm enclosure themselves. In other cases, the project has meant the women are carrying out economically productive activities. How much change this adds up to is shown somewhat more systematically by the questionnaire data.

Again, the contrast is between those in the project and other women, but one major factor must be acknowledged before any conclusions about project impacts can be drawn. ASTHA describes the participant women as 'sort of upper lower-class tribals in a poor area' because the women chosen to participate had to have land (or their husbands had to have land). This means landless women are excluded, as are women living on inadequate or particularly poor land. (Ms Varma, for example, reports on women whose land is regularly flooded and cannot grow mulberry bushes as a result). The control sample questionnaires were also not only limited to women whose husbands had 0.1 hectares of land. So the participant women are probably *de facto* better off economically than a random sample of other women in the area. With this in mind, the survey results can be considered.

All of the participants who received questionnaires lived in rural surroundings while all of the other women lived in villages in extremely rural surroundings. There was a significant age difference between the two groups; the participants were by and large older, having forty per cent in the 31 to 40-year old category and another forty per cent aged 41 to 50. In contrast, thirty-two per cent of the control women were under thirty-one.

An overwhelming majority of the women in both groups (95 and 93 per cent respectively) were married and not heads of their household. Almost all the women of both groups were Hindu; only two of the participants reported being Christian. Few of either group had a co-wife. Interestingly, more of the participant group reported not being a first wife, but the difference was not statistically meaningful. Given the younger age of the non-project women, it is not surprising that they had fewer children than the participants. Seventy per cent of the participants had two to six children, while only fifty per cent of the other women did. There was no difference between the samples in regard to education; more than ninety-five per cent of both groups had had no schooling. Significantly more participants had had literacy training but this was still only eighteen per cent of the total participant sample.

Most of both groups reported that their husbands were farmers, but a few said their husbands worked as agricultural labourers and some of the participants listed the major occupation of their husband as running a small enterprise at home. Almost twenty per cent of the other women said their husbands worked as an employee in an enterprise. There was a significant difference between the groups in regard to the education level of their husbands. Although most of both groups had had no schooling, thirty-one per cent of the participants' husbands had attended some or all of primary school or higher.

All in both groups described themselves as farmers and said they raised crops such as wheat or millet. The participants, however, were significantly more likely to earn an income from one of their economic activities (silkworm rearing and silk reeling) which most said accounted for half or more of the family income, while few of the other women made an income, much less one which was a significant proportion of the family revenue.

All of the participants had joined the project in order to earn an income. All of the participant women said they had changed their economic activities during the project period. Virtually all of them felt the project had lived up to their initial expectation, allowing them to acquire assets, increase their income, or both. Indeed, there was a significant difference between the two groups with regard to the level of family income. Predictably, the participants were more likely to be in the high-income category (none of the other women were), while thirty per cent of the non project women were in the low-income category and only fourteen per cent of the participants were.[2]

---

[2] The scale on which this distinction is based is: Rs10 000 a year equals high income; 3000 to 5000 middle; 1500 to 2500 low; and 1000 or less is below subsistence.

There was a significant discrepancy in regard to their husbands' access to assets. Most in both groups reported their husbands owning land or a house and also livestock, but where the majority of the participants said their husbands had other assets (63 per cent), eighty-nine per cent of the other women said their husbands had no such assets. None in either group, however, said their husband earned income from any of his assets. There was a significant difference between the groups in regard to their own possession of assets. Eighty-four per cent of the participants owned land or a house, while only fifty-seven per cent of the other women did – a significant disparity. Interestingly, most women in both groups reported gaining their assets during the period of the project. This, given the length of time the project had been in existence and the relatively young age of the non-project women, is perhaps not surprising. However, the participant women did credit the project itself for their relative wealth; seventy-two per cent of the participant women said the project helped them acquire the assets they had.

Both samples spent their income on food for the family. The participants, however, were more likely to use their income for household maintenance. This difference is not significant but it is reinforced by the fact that the participants did change their expenditure pattern over the time period of the project while the other women did not (and this disparity is significantly different). Moreover, many participants did report having used their income earlier for food, and not having to do so now. Sixty-one per cent of them said that the project had provided them with more money so now they could spend their income on different things.

Looking at the family life of both groups reinforces the picture of differences between them. Participant women are significantly more likely to say their families eat good food sometimes (as opposed to never) and more than ninety-three per cent of them said they ate better food now than they had before, *because* of the project. Most (67 per cent) of the participant women, however, did not feel the project had improved the chances for education of their children and, indeed, in comparing children of participant and control women, no notably different educational pattern emerges. In regard to decision-making, participant women were significantly more likely to say they alone decided about the use of their income or did so in consultation with their husbands, while more than fifty per cent of the other women said their husbands made this decision. Similarly, in regard to what they did in their work, non-project women were likely to say (61 per cent) their husbands decided this, while participant women said that they made this decision or did so with their husband's advice. Significantly more non-project women said their husbands alone decided where they lived, while participants indicated they were more likely to have a voice in this decision. Both groups of women apparently decided on the family diet. A few participants said their husbands decided on the level of schooling for their children (both girls and boys), but most women in both groups made this decision together with their husbands and the difference between the groups in regard to educational decisions was not significant.

The extent to which this difference in status measured by family authority is a result of the project is hard to estimate. Most women in both samples said the family decision had not changed during the period of the project. There were no significant differences between the groups in regard to decisions on all matters mentioned above, although the non-project women were somewhat more likely to indicate that some other family member had made decisions for them and did not now, again perhaps reflecting their youth and the presumed authority of their mothers-in-law until they had a first child and gained status in the family. Most participant women did not think the project had caused a change in the overall family decision-making pattern, although a third of them did say it had given them more overall authority than they had had. Asked directly later if the project had given them more authority in their family, however, a majority of the participants said it had. This is the same distinction which participants in other projects seem to have made. Most women did feel they had more authority in some areas as a result of the project, but did not perceive an overall shift in authority or status in the family.

Nor did most participants think the project had affected their use of time. Eighty-eight per cent said they spent the same amount of time caring for their children and their home as they had always done, only seven per cent said they spent less. Ninety-five per cent said obtaining fuel and water occupied the same amount of time it always had. Few of them responded to the question on self education and development (as presumably this is not one of their activities). But, most participant women, more than eighty-eight per cent, said they now had less time for rest or their own leisure than they had. Moreover, most of the participant women felt the project had led to a change in their time use in that they gave more time to their work as a result of it. Most of them were quite satisfied with this state of affairs, indicating that they did not want to have more time for anything in particular, although almost a third would have liked to have more time for their work (despite the fact they already spent more time on it than they had before the project). In this, they resemble women participating in projects in other countries. Although one goal shared by all donors is to lighten the load carried by poor women, many of the women themselves correctly perceive that only by working hard, and even harder, will they improve their living situations. Thus, from their point of view, a project which reduces leisure and resting time and increases work, as long as this is accompanied by increasing income, is a good and successful project.

Most of the women in the project felt it had had major impacts on their lives. Almost two-thirds said it had increased their income or allowed them to acquire more things for their family. Six women (14 per cent) noted specifically that they had attained more independence because of the project. Half of the women said the project had made them feel better about themselves while another thirty-eight per cent felt the project had improved their outlook on themselves and their economic future or both. All felt their families had profited by increased income and work opportunities and said their husbands liked the project for just these reasons. The husbands interviewed endorsed this view. Eighty per cent of the men said the project helped their family because of the increased income now available, and the other twenty per cent identified more work for more family members as the major positive impact. Three-quarters of the women could find nothing they wished changed in the project and those who did were not specific in what they wanted different. All of the women acknowledged the project's successes although they could not pinpoint just what had made it work.

Individual statements made by the women in open-ended questions show what the project meant to some of them. One said she would have had to work on the roads if it were not for the project because her husband was sick. Because of the project, however, she can work at home and care for him and her children. Another said that her husband was unable to earn money and now she could support her family. Another woman was a widow and now, because of the project, could support her family. Several participants said that because of the project they were able to stop their husbands from drinking and beating them. Overall, they said the project had given them more self-confidence and they now received more respect from their families; their husbands, and even some of their in-laws, were asking for their opinion on family matters. Some stated that they were tired because of all the new work and found it hard to get through their domestic chores. In fact, only a few said they worked less now and these were women whose families either owned or had access to a diesel pump (and did not have to haul water). But, tired as they were, the women were happy to be in the project.

In conclusion, the picture emerging here is not completely congruent with that which was suggested in the earlier evaluation of the project, where the lack of mobilization or transformation of the lives of most women involved in the project was emphasized. In contrast, the data presented here shows clear economic differences between the participants and other women. Some differences could be explained at least in part by the higher economic level of the participants at the outset, but others do not seem to be based on this pre-existing condition. Participant women have more authority in family decisions. Moreover, participant women believe the project led to changes in their lives which increased their income, allowed them to acquire assets, improved their authority in certain areas and gave them, overall, a more positive outlook. Nor is it only members of the three highly functioning Mahila Mandals reached and mobilized by

ASTHA who feel this way. Sixty-five per cent of the sampled participants do not belong to any group (six women belong to a co-operative, and another nine acknowledge belonging to a women's self-help group). Thus, the findings here reflect not only the women mobilized by ASTHA, but also the overall impacts of the TADD project in the wider group of unmobilized women. This is a very interesting result given the negative reports on the project's developmental impact in the preliminary UNIFEM evaluation discussed above. Here, after several years have passed and the project has become more internalized by an ever-larger group of people, we are uncovering quite strong developmental impacts even among women who are not in the successfully mobilized women's groups. The importance of this should not be understated. Women in the Bhil tribe have been in a very subordinate position. They do not normally speak for themselves in community matters. If they do attend meetings, they sit and listen. They 'very seldom speak; if they do so, they simply endorse views and decisions of males' (Dak, 1988:10–11). Women who do assert themselves are punished or harassed for this behaviour. Thus, in this project, 'Nirmala, a sericulture member who attended a seven-day training course outside the village, was denied entry to her house for two days' (Tawakley, 1992:7).

Women who participated in the project were clearly moving into non-traditional roles and gaining income and assets which gave them a different position in the household. In the earlier UNIFEM evaluation, the judgment was made that the women had not been able to use the income to lead to improvement in their lives, as hoped by the project planners. Yet, there may be some question about this as a total judgment. We have no evidence that this report was wrong when it stated that food shortages and increased malnutrition and ill health resulted from planting mulberry trees in place of traditional food crops. This was a negative if, in fact, there was not a general food shortage in the zone experienced by non-project families as well. Other criticism, however, may not be as warranted. The report cites increased alcohol consumption by both men and women and the purchasing by the women of non-essential items such as jewellery or household utensils with their profits. The latter is a strange criticism because, in this zone, women do not usually own property or have savings accounts. Women's traditional assets *are* jewellery and household utensils which may be converted to cash or used as collateral for a loan in times of need. Within Bhil customary law only men can inherit land and, within the Bhil tribe, the wife of the deceased husband can be asked by her sons to leave the family home. With inequality in wages and lack of opportunities for jobs in the district, a woman can not expect to earn an adequate income to support her family in times of crisis. Women's assets in the form of jewellery may serve as financial insurance to sustain them, at least temporarily in hard times. The devotion of their profits to this end is not only understandable but, in the cultural circumstances, perhaps advisable. Both men and women acknowledge that jewellery belongs to the women. Husbands who answered this questionnaire noted that they owned houses, livestock, machinery and even savings accounts, but their wives had jewellery. A savings account might be seen as too businesslike and not womanly, whereas jewellery does not (Rodriguez-Streeter, 1994). Thus, it is no surprise that in this questionnaire a *positive* comparison between participant and non-participant women was the amount of silver jewellery owned (while few women had savings accounts even after the intervention of ASTHA). Both groups owned some jewellery but there was a significant difference in favour of the participants. These women owned more jewellery, worth at least as much as their family's annual income (which was higher on average among the participants, as noted above). This measure of success (as seen by the women participants) is a standard of women in other cultures with similar constraints: jewellery is a form of savings for many.[3]

A further point of the earlier evaluation was that men were inserting themselves into the project and taking over roles that women had been expected to play. Here again, this was seen as negative but perhaps some care should be taken with this conclusion. For the women, this involvement by their husbands did not mean that they had no role or that they got nothing from the project. A woman's husband's acceptance may have been facilitated by his involvement. Again, just as women in other situations found their new importance made their work become an object of interest to their husbands (see Ghana discussion later), in this case men insisted on some part in this new and important family undertaking. The women did not necessarily see this as bad, and, indeed, it may have been a positive factor because it made the project more acceptable to the men. In fact, women gained in authority over some matters (see above) and gained in income. Perhaps this involvement of the men was natural in the process of bringing women in.

This discussion is not meant to imply that everything for the Bhil women participants was perfect. Clearly, there is much more progress to be made if these women are to gain more equal status, but the possible impact of relatively long-term access to regular income among women who had not had this opportunity before, even without intensive mobilization in women's groups, is quite strongly demonstrated. As in the case of Guatemala, where the men's economic success resulted in improved conditions for their wives, this finding does not suggest to us that mobilization of women on a broad scale is not needed. Women in the mobilized Mahila Mandals had a new role in the community and a new sense of themselves which is not reflected in the women who were not in the mobilized villages.

---

[3] For example, women participants in Senegal in a SME project used their jewellery as collateral to get a loan from the project and bought jewellery with their profits (gold in this case). See Vengroff and Creevey, (1994).

These results do show, however, that successful long-term income-generating schemes will have developmental impacts reaching beyond a mere increase in revenues available.

Assessing the record

The Women's Sericulture Project has definitely met the goals it established for itself at the outset. Employment opportunities have been created in a zone where virtually no alternatives existed outside of traditional agriculture, and which has been losing ground owing to worsening soil and climate conditions. Family incomes have been substantially improved for those families participating in the project (see Annex 3 for calculations of profits earned by type production).

Nor have the benefits been restricted to a small group of participants who happened to be chosen for the original project. TADD has gone beyond the five hundred official participants and works with an additional 2500, all in the zone. The end point of this expansion may have been reached, however, and the future is not completely clear. The market for the cocoons and reeled silk seems to continue to be without limits but the profit margin for most family productions is very small. The World Bank programme, which is introducing a more sophisticated technology, should realize higher profit margins and greater output, although the workers in their programme will be largely hired staff, probably with low wages. The Rajasthan Sericulture Project may have to re-educate its participants and introduce new, more expensive techniques. This would add to the cost of the project as subsidies would be needed. An intensive study of the feasibility of altered technologies of production in this zone and their impact is needed.

The project has had an impact on the role of some of the women in the villages and has given them new responsibilities and new authority. This change has gone beyond the economic sector so that, in a few cases, women are becoming influential in local politics, something which had not been common before. Although there are indications that in some Mahila Mandals women have been made more aware of their own potential and have established businesses and undertaken other community activities as a result, this is not widespread. Only three of the Mahila Mandals have been extremely active. Three others are making slower steps forward. There are no organized women's groups in the other six project villages. ASTHA has done an excellent job in working with some of the village women, but its progress is nonetheless slow and has touched few people. Women are hesitant to put themselves forward and many do not understand exactly what ASTHA is purporting to teach them. Some feel that the Mahila Mandals are a waste of time because they do not see the connection between these groups and their sericulture work. The inability of ASTHA and TADD to collaborate impedes the realistic growth of the role of the Mahila Mandals. Also, ASTHA is clearly understaffed, having only one person to work with all the groups. This is a substantial barrier to mobilizing and sustaining the activities of ever-larger groups of women.

Finally, the women in this project continue to be under the control and domination of their male relatives. Among the Bhil, women have more freedom than in many parts of India – they may, for example, seek another husband when they are widowed. However, their situation is still one of subordination. Men own the land on which the trees are grown and have the overall direction of the household earnings, including those from the silkworm project. Women are increasing in power, especially where the women's group is strong, but it will be a long time before there is any serious revision in thinking about the role and importance of women in the community at large.

There is still a question as to the impact of the project on society in the villages as a whole. The project participants are poor in a severely impoverished area. However, they are not the poorest of the poor or the most disadvantaged of their society (as were the participants in Bangladesh for example). In this case, the project works with women whose families have land and a well for water for the mulberry bushes. Landless women, who are certainly the poorest, are excluded. Thus as quoted above, one ASTHA staff member said in interview that ASTHA works with the poor but this project, in contrast, works in a poor area with 'sort of upper lower-class tribals in a poor area'. It is understandable and reasonable that some minimum criteria were adopted for choice of participants; without land or water, a woman could not raise the bushes or the worms necessary. It may be necessary to go back to the villages with this specific factor in mind to see the impact of this new opportunity on the social and political order of the villages. The green revolution has often been criticized for creating rural elites and dispossessing the non-elite. This project should not follow the same path. Perhaps it might be possible to develop a sub-project which would directly aim at drawing in women who could not meet these economic criteria but who could be specifically included in group economic activities.

An additional consideration is the inter-agency collaboration exhibited by this project. Not only is the successful collaboration among many varied agencies a useful precedent, but the fact that the major implementation agency is a governmental agency is an important factor. TADD is responsible for the government's work among the Bhil. TADD will continue to be responsible when the project stops and outside funding diminishes or simply ends. It can continue, if necessary, with no additional funds from UNIFEM. Given that continued reforms and revised plans for the sericulture production may be required in the future, and that the participants will need extension and oversight for many years to come, as well as technical backstopping, it is significant that the enduring responsible agent is fully involved and aware. It is also relevant that UNIFEM and TADD were able to work closely together and that UNIFEM was able to monitor and periodically evaluate the progress of the project. There did not seem to be any major conflict despite the withdrawal of UNIFEM at the end of the first phase. TADD was able to accept the need for a second stage, which included ASTHA and an emphasis on mobilizing and training women to manage their own development process. This outcome suggests good collaborative procedures, and a tolerant and co-operative staff, which sets an excellent precedent. Not all problems in collaboration have disappeared, but the overall record is quite good.

A final question is the degree of dependency of the participants on TADD and the other agencies. The project was conceived and developed by a government agency and outside donors; there was little participant involvement in decisions which were made. The technology and feasibility were tested by academic researchers and a local NGO. Farmers were persuaded to join by subsidies and promises. The organization of the activities was determined by TADD staff and 'taught' to the participants. Until the arrival of ASTHA, the women who were the targets of action were not decision-makers and were, therefore, not really responsible for their own activities. Even now, most of them do not take a strong or independent role in economic decision-making. All marketing is still done through TADD. The project has demonstrated that women can learn how to grow mulberry trees, raise silkworms, reel silk, etc., but it has not shown that they or their families could independently conduct the enterprise. Furthermore, ASTHA continues to be dependent on funds from UNIFEM or some other outside source. Its work will not continue unless such funds are available and, without continuing support over a fairly long period, it is doubtful that women will be mobilized to improve their positions in society or take any more responsibility. One TADD official said in an interview that the Bhil are uneducated and untrained and 'have no economic conscience'. What is clearly meant is that they do not have a full grasp of what they are engaged in. And, they have had few needs or exposure to wealthier individuals with higher standards of living and thus, less incentive to work for larger incomes than others who have been exposed to the possible goods and services which could be purchased. This may change over time. Perhaps it would be unrealistic to expect that such an 'economic' attitude would develop among most people very quickly, but it remains to be seen whether it will result from this project over the long run. Without it, this can not be a self-sustaining, independent set of activities which must be a major goal of this effort.

So far, however, this project has had major impacts on the lives, status, time use and attitudes of the women who have participated in it. These are substantial results and they are the more interesting because a large number of women have been affected. This is not just fifty or a couple of hundred but rather thousands of women, and the enterprise is still being expanded to include even more women in the zone. Despite the long-term problems of competition, profit margins and continuing dependency, this project is one of the few which can be said to have had a major developmental effect in the region where it has taken place.

Chapter Six: Statistically Significant Survey Analysis Results in India[4]

1) Sample type (participant, control) and age (15–20,21–30,31–40,41–50,51–60)
   Chi sq. = 10.298, Cramer's V = .381, P. = .0357
2) Sample type (participant, control) and literacy training (none, some)
   Chi sq. = 5.038, Phi = .281, P. = .0248
3) Sample type (participant, control) and husband's education (none, part primary, all primary, part secondary, all secondary)
   Chi sq = 7.562, Cramer's V = .341, P. = .056
4) Sample type (participant, control) and family income (high, middle, low, below subsistence)
   Chi sq. = 12.501, Cramer's V = .423, P. = .0019
5) Sample type (participant, control) and husband's ownership of other assets (none, few,)
   Chi sq. = 17.319, Phi = .501, P. = .0001
6) Sample type (participant, control) and respondent's ownership of land/house (none, some)
   Chi sq. = 3.502, Phi = .222, P. = .0613
7) Sample type (participant, control) and respondent's ownership of other assets (none, some)
   Chi sq. = 6.112, Phi = .293, P. = .0134
8) Sample type (participant, control) and respondent's receipt of income from primary economic activity (half or more of family income, small contribution to family income, small and insignificant income, no cash income)
   Chi sq. = 35.961, Cramer's V = .787, P. = .0001
9) Sample type (participant, control) and access to luxury food items (rarely or never)
   Chi sq. = 5.024, Phi = .27, P. = .025
10) Sample type (participant, control) and change in expenditure of income during project period (no, yes)
    Chi sq. = 11.3, Phi = .424, P. = .0008

Decision-making

Type of decision	she decides		husband		family member		husband & wife	
	P	C	P	C	P	C	P	C
Use of income	11(26%)	3(11%)	13(30%)	14(50%)	0	1(4%)	19(44%)	10(36%)
Chi sq. = 5.477, Cramer's V = .268, P. = .14								
Decision on her work	15(35%)	2(7%)	12(28%)	17(61%)	0	1(4%)	16(37%)	8(29%)
Chi sq. = 11.829, Cramer's V = .408, P. = .008								
Decision on dwelling	5(12%)	0	24(59%)	24(86%)	0	1(4%)	12(29%)	3(11%)
Chi sq. = no results (because of lack of data in third column)								

P = participant, C = Control

---

[4] The results of the decision-making comparisons are also reported here, including both the percentages and the chi square results because of their discussion in the text.

CHAPTER SEVEN
Thailand – Venture Capital and the Pickled Ginger Enterprise

The development miracle

Thailand is a country with 55.8 million people living in a land area of 1 139 000 km^2. In contrast to its Asian neighbours discussed here, the average annual per capita income of the Thai population is US$1420. The more relevant context, however, is the growth of the economy. Where the GDP of India grew 3.6 per cent between 1965 and 1980 and 5.3 per cent from 1980 to 1990 (more than any other country yet discussed), Thailand's economy far outstripped this performance. In Thailand, the GDP grew 7.3 and 7.6 per cent respectively in the two periods. And the industrial sector established itself, providing, in 1990, 39 per cent of the GDP, while agriculture went from being 32 per cent of the GDP in 1965 to only 12 per cent in 1990. Inflation is not a problem in Thailand. The inflation rate from 1980 to 1990 was only 3.4 per cent, less than that of the United States in the same period. The population of Thailand was rapidly moving into the cities in this period, going from only 13 per cent urban in 1965 to 23 per cent in 1990, with the large majority of these being in Bangkok (World Bank, 1992:218, 220, 222, 278).

Despite the positive national economic performance and the transformation of the structure of the economy, much of the rural population of Thailand, especially in areas less accessible to the capital region, faces very difficult economic conditions. As much as 28 per cent of the rural population lived below the poverty line in 1989 (Heyzer, 1989:1110). The region of the project discussed here, the north and north-east, is one of the least favoured areas. Where the average *per annum* GDP per capita is $1300 for the country as a whole, here it is $425. Forty-eight per cent of the population in this region live in households whose income falls below the poverty line (Hyman, Gupta and Dayal, 1993:5).

Certain factors, however, augur well for improvements in this situation even in the north and the northeast. One of these is the decrease in birthrate, which has gone from 7.0 in 1965 to 3.6 in 1990, the lowest of any of the countries studied here. Certainly related to this is the spread of education. By 1989, 86 per cent of the relevant school-age population in

Thailand was in primary school, which, although a lower figure than those of India or Peru, still indicates that the majority of younger residents have this option, particularly in towns or cities but increasingly in the countryside.

Quiet power: Thai women

Thailand is a Buddhist country. Thai law traditionally permitted polygamy and men were seen as the head of household with authority over the women in their families. Male domination, however, was more evident in the upper and wealthier classes, especially in the urban areas and in the southern rural zone. In the rural central zone and in northern Thailand, including the project area, women had distinct and recognized authority in the majority of families. Residence was primarily matrilocal, that is, a married couple settled with or near the wife's family. Legal authority over the family possessions was in the hands of men (passing from father-in-law to son-in-law in the north-east region, for example). But women were significant decision-makers. As Chai Podhisita has written, 'Here, women are the key members of the household through whom the majority of the household is obtained and regulated' (Pongsapich, 1991:258).

The tradition of a certain amount of female authority, and the presence of women in public places such as the market-place and city streets gives Thai women some advantage over females in Bangladesh and India. This relative independence is reflected in both their educational and economic positions in society. Only 3.8 per cent of the female population aged 15 to 24 is illiterate, as compared to 2.4 per cent of males. In the older age group, 23 per cent of women, as compared to ten per cent of men, are illiterate but this difference is being eliminated by schooling patterns. The ratio of female to male school enrolment shows that, in 1980 to 1984, girl's enrolment in primary school was 93 per cent of the figure for males. These educational figures clearly demonstrate not only that most women have the possibility of access to at least some level of schooling, but also that there is no significant difference between their access and that of their male age cohorts. This is a very different situation from that of most of the other cases discussed so far (UNIFEM, 1991:53).

In the Thai economy, positions for women have opened up at a more rapid rate than in most other developing countries. Rural Thai women have traditionally been fully occupied in the family farm activities, growing vegetables, raising swine and poultry, etc. Far more than in most poor countries, their activities are recognized as economically productive. Women are, in fact, 45 per cent of those cited as economically active. Forty-six per cent of all women are seen as economically active (the highest figure for any Asian country except China). Of these, 77 per cent are engaged in agriculture, 13 per cent in the services sector and 15 per cent in the industrial sector. In agriculture and related forestry activities, there is approximately equal male and female representation in the overall workforce. In the wage sector, they dominate clerical and service positions, outnumbering men substantially. Perhaps even more striking, in the mid-seventies, after large industries had begun to relocate to relatively low-wage countries, Thailand was one of the beneficiaries. Many multinational and foreign national concerns opened in Thailand and hired women as their line workers. However much the working conditions of these export zone businesses may be criticized, women flocked to these jobs perceiving (correctly in purely monetary terms) that the salaries were better than what was otherwise available. As a result, women in Thailand had become 44 per cent of the industrial workforce by 1984. In addition, Thai women commonly have assets of their own such as land, buildings, savings and investments, or even businesses. This is reflected in the Thai Government's Household Accounts, where not only does the government list family income and family assets, but distinguishes between male and female assets in the family, listing substantial assets for women in all areas and even listing assets for women among the poorer sections of the population (Heyzer, 1989:1110, 1117; UNIFEM, 1991:107).

Compared to most of the other cases we are evaluating in this study, then, Thai women are faced with a reasonably open and growing economy with increasing alternatives for jobs, especially in the urban areas, without the same restrictions on their activities. This situation is not quite as rosy in remote rural areas, but the tradition of women being involved in agriculture and agro-businesses is well established even here, creating a more favourable climate for women to establish small enterprises than in other, more restricted, societies with more limited economies. Thai women, however, believe that the situation can and should be dramatically improved. They point out that women are of lower status and in the worst-paid positions. Women are 65 per cent of the unpaid family workers, for example (UNIFEM, 1991:111). Nor has the Government of Thailand been willing to ensure or enforce equality for women in the market place. For a brief period, the 1974 constitution guaranteed the end to all discrimination, but this provision was dropped in the succeeding constitution. In the recent unrest of 1991, which was followed by an Interim Government in 1991 and finally an elected democratic regime in 1992, no major efforts were made to address the needs of women; the government remained relatively conservative in outlook. Small enterprise development and programmes for rural income generation are supported by the government. In addition, a variety of government programmes and NGO projects have attempted to improve matters for Thai women, offering loans and technical support for economic activities, but a disparity between the rights and authority of males and females still exists.[1]

The Joint Venture Capital Project in Northern Thailand

The rationale for the Thailand project is best explained by the following:

> The purpose of a venture capital project is to overcome a market failure in the financial system of less developed countries. Although rural enterprises involving small-scale producers can often be profitable, it is usually difficult for them to start operations. The major constraints are the inability to meet strict collateral requirements for credit and the need for technical and managerial assistance. From the perspective of the formal sector financial institutions, such enterprises are risky and the transaction costs of dealing with them are high. Even in the United States, the failure rate for new businesses (and not just those in underdeveloped areas) is 75–80 per cent. Expectations for the survival rates of new businesses in less developed countries where constraints are even more severe in many ways need to be considered with this perspective (Hyman, Gupta and Dayal, 1993:1)

What ATI proposed to do in this situation was to support a local NGO, the Population and Community Development Association (PDA), in establishing a company which could provide capital for small and medium-scale enterprise formation. The original stated objective (Gupta, 1992) and continuing goal of the new company, called Rural Small Scale Industry (RSSI), was to support and promote 'grassroots corporations as a new force in the economic growth of the nation by developing an outlet for technical and entrepreneurial skills that otherwise might not be fully utilized'. This venture capital company would provide financing and other technical and managerial assistance to businesses which local entrepreneurs were attempting to set up. The businesses would be chosen on the basis of the prospects for commercial viability, use of appropriate technologies, and social benefits. These enterprises could be owned and managed by individuals or groups, but the 'initiative was to remain in the private sector'. The venture capital company would provide a portion of the initial equity, that is, would initially own a portion of the enterprise,

---

[1] In recent years, a variety of women's groups have sprung up which are beginning to challenge this position of inequality. Many of these started on university campuses; others were allied with religious groups. Some of them include The Association for Civil Liberty, The Christian Women's Association, The Women's Information Center, the Child's Rights Protection Center as well as many others. These groups, largely formed by educated women, focus on two major issues: bringing about equal rights for women and ending child labour and prostitution. So far neither goal has been reached (Pongsapich, 1991:260–66).

but other sources of equity and/or loan capital were to be leveraged as a result of the financing and other assistance provided by the division (Gupta, 1993:1).

PDA established RSSI in 1984. In each case, where an enterprise was supported, RSSI would provide equity capital and receive a proportion of the profits made by the enterprise equal to the share of the equity capital advanced. In other words, RSSI became a part-owner until the company could buy out its shares. This system is seen as preferable to simple loans which impose an obligation of both interest and debt repayment whether or not the company is successful. In contrast, RSSI's equity capital was exposed to the same risk as the capital of the local entrepreneurs. In the case of a failure, this capital may be lost. Where the business is successful, the local entrepreneurs will buy out RSSI's share and that money (and whatever profits engendered by it and received by RSSI), can be invested in another enterprise. Venture capital companies (and RSSI) also charge fees for the specific technical and managerial assistance they provide. Charges for these services and profit sharing on the basis of equity provided are part of the attempt to assure that the venture capital companies' costs will be covered and that increasing resources will be available for future enterprises, which will also be supported. Exiting from the new enterprise once it is capable of proceeding on its own, or has proven that it is not viable, is standard so that the venture capital company can either then recycle the money to a new company or cut its losses. The objective of the procedures adopted is to create a sustainable financial institution which can continue over the long term to supplement traditional lending institutions in providing funds to assist small entrepreneurs.

RSSI was registered in November 1984 with a share capital of $320,000.[2] One of the enterprises it initially supported, which is the main subject of this analysis, is the Chiang Rai Agro Industry, Inc. (CTA).[3] In December 1988, Mr Akraphole Vatcharakomolphan, a 47-year old trader and processor of fruits and vegetables, approached RSSI for assistance. He wished to establish a company for fruit and vegetable processing and distribution in Chiang Rai, the northernmost province of Thailand (Chiang Rai, 1988). RSSI asked the Department of Industrial Promotion to conduct a feasibility study on this endeavour. This study suggested that CTA should focus on the production of ginger, which showed excellent promise for a favourable return on investment.

Ginger grows well all over Thailand, especially in the north where 47 per cent of the ginger production in the country takes place. In 1981, Thailand exported 4042 tons of pickled ginger and, by 1982, the volume and value of

---

[2] The current exchange rate is 25 baht to $1.
[3] Two other investment projects were approved by ATI and financed by RSSI before CTA.

these exports had more than doubled. As a result, the area planted with ginger in Thailand doubled between 1982 and 1987 from 25 654 rais to 46 911.[4] Ginger is used in cooking, in medicines and in drinks. Its growing season lasts from March to June. Young ginger is harvested after about four and a half to five months and sold largely for export. Older ginger is harvested later and sold and consumed locally. Farmers, however, can not grow ginger continuously due to the prevalence of plant diseases, but must give the land a two-year rest between ginger harvests. Usually, Thai farmers leave the land fallow in this period as there is enough land for crop rotation, but, if they have used adequate fertilizers, they can grow other crops such as chillies or tomatoes in between ginger harvests. The major foreign market for the ginger produced in Thailand is Japan which bought 90 per cent of Thailand's ginger in 1990 (Hyman, 1992).

Following the receipt of the feasibility study RSSI agreed to assist in the formation of CTA. Initially there were seven individual shareholders, of whom four were women and three were men, who provided five million baht. RSSI provided one million, and a local bank, the Siam Commercial Bank, loaned a further 6.5 million baht at 13.5 per cent interest per annum. CTA would purchase ginger from the local farmers and hire workers for the processing, which included washing and peeling by machine, fermentation in a solution of salt and vinegar, washing again to remove impurities, polishing, size sorting, packing and distribution. An agreement had been reached before the financing was finally made available between RSSI and the Tokyo Trading Company, Ltd so that the latter would purchase CTA's product. In return for this monopoly, the Tokyo Trading Company provided technical assistance in ginger-processing methods and factory administration (Traitongyoo, 1989).

Beneficiaries from this enterprise would not only be the entrepreneurs but also the employees who, at the height of the season, would number 300 to 400 and would be primarily women. The farmers, including both men and women, who sold ginger would also profit from their increased market. They were linked to specific investors in CTA who were all traders and had their own network of farmers from whom they bought local produce. As each trader was 'linked' to about 100 farmers, this meant that the pool of beneficiaries from this project (although not all simultaneously) would be as many as 700 farmers. RSSI favoured the business established by the CTA because of its impacts on a region which is one of the poorest in Thailand. In particular, it supported the increase in employment opportunities it would provide through factory jobs, the increased income available to ginger growers, and the ability chance to ginger growers to increase their bargaining power in the marketplace.

---

[4] A rai is approximately an acre.

Venture capital in action

The Chiang Rai Thai Agro Industry Co., Ltd (CTA) was established in January 1989 in Viang Pha Pao. It was supported by the Thai Board of Investment which allowed it exemption from corporate income taxes for eight fiscal years from the date that income began to be earned, from import duty, and from business tax on machinery, component parts and accessories. The investor who had initiated the project, Mr Akraphole Vatcharakomolphan, was the initial managing director. A second investor, Ms Jitpranee Huntanon, who owned a shop selling construction materials, managed the necessary building for the project's factory and offices. RSSI provided two permanent staff, hired from Bangkok, for technical support in management and marketing.

The original shareholders had little experience of running a company and there was a certain friction or lack of trust among them. In these circumstances, RSSI's role in technical backstopping was exceptionally important. One major issue where RSSI intervened was when, in the first year, the original shareholders were considering selling their shares to the Tokyo Trading Company, which held the rights to buy all of their product and was controlling the price they received for the pickled ginger. RSSI wished to keep the CTA a local company with the potential for improving the lot of farmers, traders and investors (in line with the original mission of PDA). It therefore purchased 5000 more shares in 1990, bringing its equity in CTA to one and a half million baht.[5] The General Manager of CTA thereafter decided to sell the ginger to whomever paid the best price and the original agreement with the Tokyo Trading Company was abrogated (although the Tokyo Company could still compete to buy the product).

CTA also faced a problem in finding labourers for the factory as the peak period coincided with a peak period for agricultural work in the fields. This shortage of workers affected the company's capacity to meet sales demands. During the years in which the CTA produced ginger, they did not completely resolve this problem and, both because of this and because of the fluctuating needs of the production process itself, the numbers of employees rose and fell at different times of year. In September 1990, for example, 130 females and 40 males were working in the factory. In June 1992, the factory employed 450 workers, 400 of whom were women (Hyman, 1992:2). During the off-growing season for ginger, January to June, the factory was idle.

A second initial problem involved cash flow; because of the terms of the letter of credit from the Siam Bank, CTA did not have sufficient working capital to buy enough young ginger for processing. RSSI and the Siam Bank helped CTA overcome this difficulty.

---

[5] This additional money was not provided by ATI but came from general RSSI funds.

Originally, pickled ginger from the CTA was to be exported in wooden boxes to Japan. Shortly after the project started, however, the price of wood rose steeply. One of the original investors had an associate who was a potter and he began to produce ceramic pots to contain the ginger. These pots were very popular in the market in Japan and the factory switched completely and successfully to this form of packaging, thus stimulating the local pottery market (Hyman, 1992:2).

In February 1990, Ms Jitpranee Huntanon bought out the shares of her fellow investors except for RSSI. She was now the managing director. In September of the same year, RSSI estimated that the CTA could manage on its own and sold its shares to her, and Ms. Huntanon, together with her son, became the sole owners. At this point, 16 months after the CTA had been started, the shares had appreciated in value from 100–120 baht a share. The RSSI had made a substantial profit on its investment (300 000 baht), as had the original investors. Ms Jitpranee thereafter sold 65 per cent of her shares in CTA to the Thai Choo Ros company from Bangkok. The new managing director was appointed by Thai Choo Ros.

In 1993, the price of young ginger had risen sharply as a result of the fallow period for ginger throughout the zone. In addition, a Japanese ginger pickling unit entered the market for the young ginger, thus driving the price up further. CTA decided not to produce ginger for the time being. From January to May 1993, the factory was rented to a cabbage pickling enterprise (SANTIPARB) for storage space. CTA's future plans, either for agro-processing of ginger or other fruits and vegetables, or other economic activities such as rental, were unknown at that time.

Impacts and changes in individual lives

The overall objective of this study is to assess the impacts of different types of strategies on the women who are involved as participants, or on women in the family of male participants. In the case of the Thai Venture Capital Project, it is necessary to distinguish carefully between the impacts which may have taken place on four types of participants: the investors, the traders, the farmers who sold ginger, and the women who worked in the ginger pickling plant. The last two groups were canvassed through the use of the same questionnaires used for the other countries studied here. The first two were interviewed by Rachitta Na Pattaling using a questionnaire adapted from the enterprise questionnaire by Valerie Autissier.

The investors as a group were relatively wealthy people already, with numerous business interests before the project started. Four investors responded to questionnaires, two men and two women. They ranged in age from 39 to 61 years old, all of them at least originally from the local town of Vieng Pa Pao. Both of the women had completed grade four while one man had a university education and the other had finished secondary school. All of them are successful entrepreneurs; one of the women is described by Ms Pattaling as a 'millionaire'. Their businesses are extensive and varied: they are all investors in more than one activity. The older and wealthier woman has a row of houses which she rents out, as well as a rice mill and a rice farm. The younger woman has a food-processing factory which she opened after she took her money out of CTA. One of the men is the son of Ms Jitpranee, the original investor, and he has a shop selling construction materials. The other male investor has a food-processing factory and also acts as a trader for food crops in the area.

Although the comments of these four vary, certain major points emerge. First, all of the four seem relatively unaffected by the end of CTA as a ginger-processing unit. They withdrew their capital and profits and invested them in other things. Secondly, none sees the CTA as having a major influence on them, even on their incomes, although they generally acknowledge that, although it did not issue dividends, they did make a profit on their shares. Thirdly, all were glad of the opportunity to invest their money and praised the CTA for its efforts at rural development. In addition, three of them said they had learned from being involved in the project and that they were using their knowledge in their current investments. Only Ms Jitpranee's son had no comment to make other than that he knew nothing about CTA and had only invested because his mother told him to. Interviews with Ms Jitpranee, in addition, reinforce this picture. Although she ended up unable to continue the ginger business, and the factory, now used as storage, is less profitable, she gained from her sale to Thai Choo Ros and is happy with the opportunities the original investment offered her.

Ms Pattaling interviewed five traders who had dealt with CTA, and four who had not dealt with them directly although they knew of it. The CTA traders were relatively young, ranging in age from 39 to 46; three were men and two women. The men had completed fourth grade, but one of the women had finished seventh grade and the other tenth. None traded only in ginger. One of the women has a fruit canning factory processing such things as bamboo shoots, mango and lychee. She grows ginger on her own land and trades in numerous crops. She has a staff of four and several other daily workers. She is the wife of one of the large shareholders in CTA. Another female trader has fewer assets and is primarily a trader in ginger, corn, cabbage, cucumber and bamboo shoots. She has no permanent staff but hires daily workers to pack and load trucks. One of the men traders is a construction worker as well as a corn and rice farmer (on his own land) and trades whatever crops are in season (such as ginger, corn, etc.). He and his wife do most of the work together. The second sells construction materials, grows ginger himself, provides transportation services, and trades in ginger. He has two full-time drivers and ten agricultural workers in the appropriate growing and harvesting seasons. He also bought shares in CTA, about which he heard from one of the original large investors. The third trader is also a rice and ginger farmer and trades in many different food crops. He is considered a 'junior trader' and distributes his crops through senior or more large-scale traders. Nonetheless, he has as many as thirty people helping in packaging and doing manual labour.

Most of the traders did not indicate that the demise of the CTA ginger-pickling enterprise had seriously affected them. One of the female traders said that she had had to go into selling religious goods because she had lost the ginger market outlet which CTA provided. Another female trader said it did not have much impact on her, although it had been convenient to sell to CTA because the factory was so close. The men traders reported that they were not very affected by the closure of the enterprise. One said he had considered selling to CTA but the price offered was too low, so he sold to someone else. Another just commented that conditions changed all the time anyway and traders had to adjust to whatever the market dictated. Most of the traders did not know much about CTA except that they had once sold to it. A couple offered negative comments such as 'it would have been better for the farmers and traders in the area if they had maintained the enterprise longer', or 'they should not have settled a factory in a residential area and should have treated the waste water from the plant'. Additionally, two commented that 'CTA should have had a fairer pricing system'. These comments, however, did not indicate any major impact on their own economic livelihood and they were apparently happy to deal with the company when it was processing ginger.

The interviews suggest that the CTA experience did not have the kind of developmental impact on investors or traders which we have explored in other projects. These people took advantage of the economic opportunity offered by CTA. They profited from it and some learned something from it to add to their business experience. Their work and their lives were not altered by participation. Nor were they personally disadvantaged in any major way by the disappearance of the ginger enterprise, they just moved on to other opportunities. The overall impression from the interviews is that of a region with on-going activities in all kinds of food processing and other areas (such as construction), and a group of entrepreneurs who already knew before CTA how to use the openings offered them. Although there is no direct income measure for the investors and traders, they obviously were quite economically successful relative to the average standard of living in the region. Their CTA experience contributed to their income during its existence. They now look to other alternatives, but philosophically, as the market is always fluctuating.

The ginger farmers and workers in the factory are much poorer than the traders and investors interviewed. The women who worked in the factory knew directly about CTA and might have had some reaction to its influence on their lives, although they worked for only an intermittent (seasonal) and short time. The farmers, at least, had a reaction, in terms of the impact of losing this particular marketing outlet and also losing whatever assistance traders had given them during the project (three traders mentioned that they had supplied fertilizer and selected ginger to the farmers but did so no longer).

There were some negative reactions to the factory in terms of its environment and health impacts, both from neighbours and from workers, which Ms Pattaling records in her notes. Neighbours complained that the factory 'smelled.' Water treatment, mentioned by the traders, was also mentioned by neighbours who complained that the untreated wastewater from the plant ruined their rice fields. One of the workers complained that the salt and acid used in the pickling made her sick. On the other hand, Ms Pattaling reports that most of the workers were satisfied with the wages they had received while working in the factory.

Overall impacts measured in comparisons of questionnaires administered to those who had sold to or worked for CTA, and those who had no involvement with CTA, were quite interesting. The sample of participants[6] (farmers and workers) included thirty-six women and there was a sample of nineteen women drawn from those who had had no contact with the ginger pickling factory. Most of the women (over seventy per cent in both samples) lived in rural areas rather than in a village or town. The participant women were somewhat younger than the other women; sixty-seven per cent of them were twenty-one to forty years of age and only thirty-seven per cent of the other women were in this bracket, forty-two per cent of them falling in the over-forty group. None of the women in either group said they were the head of their household. Almost all of both groups were married and had between one and four children. Educational patterns were also similar. Over eighty-four per cent of both groups had been to some or all of primary school while a slightly larger percentage of the uninvolved women had attended some or all of secondary school. On the other hand, only among the participants had any of the unschooled women experienced literacy training, but the number was small and the difference between the samples not significant. All of the women were Buddhist, and most women said their husbands were farmers. Eighty per cent or more of both groups of women said their husbands had attended some or all of primary school.

In regard to their view of their own occupational status, most women in both groups said they were housewives, but the uninvolved women were more likely to identify other primary designations such as running a small enterprise at home. More than eighty per cent of both groups reported they

---

[6] For simplicity, those who sold to or worked in the factory are called 'participants', although they were not actually participating in a project as such.

grew crops such as corn or rice (despite classifying themselves as 'housewives') with only a few indicating that they grew ginger. A majority of both groups of women reported selling what they grew, although from thirty to forty-five per cent said they used what they grew for their families' needs. One distinct difference between the samples was that a significantly larger number of participant women worked as wage labourers in agriculture. Women who had not sold to or worked for the factory were also different from participants in their involvement in economic groups and cooperatives. Although most women in both groups belonged to a cooperative, women who had no contact with the factory were significantly more likely to be in one. They were also more likely to be in some other women's economic group, although this difference was not significant.

Ninety-one per cent of the participants had joined the project (in the sense of selling to it or working in the factory) for three to five years. Most said they had become involved because it offered them a higher income than they got at the time. Thus, they benefited from an increased income during the period they worked at the factory (or sold to it) but, of course, this ended when the factory closed. The project, therefore, had no long-term impact on their income. Most of the women said the project had had no impact on their economic activities and/or did not increase their income overall. Nor was there a significant difference between the uninvolved and the participant women in terms of family income. Most women in both groups (almost seventy per cent) fell in the middle-income group (on a scale where 50 000 to 60 000 baht equaled middle income).

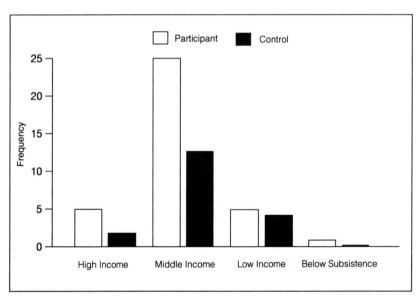

Figure 7.1 *Income levels*

Looking at the ownership of assets of the two groups, most women said their husbands did not own property or a house, did not have cattle and did not have other assets. A few more of the women participants said their husbands owned a house or other assets, but the differences in the samples were not significant. In both cases, also, most women said their husbands received no income from their assets. In terms of their own acquisition of assets, however, the situation was reversed. Although most women in both groups did not have land or a house or livestock, but did have other assets, uninvolved women were *more* likely to have land or a house, just as likely to have livestock, and more likely to own other assets. Neither group received an income from their assets of whatever kind. Nor were these assets owing to the project for most of the participants. Only seven women (20 per cent) said the project helped them acquire assets. Indeed, the uninvolved women were more likely to have acquired their assets during the period of the project than the participants were!

The participant women did acknowledge some changes in their lives over the period of the project. Fifty-six per cent said they spent less time working in their household and caring for their children. More than a third said they spent less time gathering fuel or obtaining water, but fifty per cent said these activities had not changed. Fifty per cent said they had less time for leisure or rest than they had had, but almost forty per cent said this had not changed. A few women (15 per cent) said they spent more time on their own education and self-development than they had before, while more than seventy per cent said there had been no change. Asked if the project was responsible for the changes in their use of time, fifty-six per cent said 'no', while those who said it had an impact said simply that it had led them to engage in different economic activities. Nor did most women in both groups wish to change how they used their time, although fifteen per cent of the participants said they wished they had more time for their family.

All women in both groups spent most of their income obtaining food for their family. This had not changed during the project period, and none of the participants felt the project had any effect on this expenditure pattern.

In regard to other measures showing differences between participants' and other women's family lives, over 97 per cent of the women in both groups said they ate good and expensive food regularly. Eighty-two per cent of the participants said their family diet had not been affected during the project period. No one thought the project had had any impact on the family diet at all. More than ninety per cent of the women thought the project had had no effect on the education of their children and, indeed, no differences emerge in a comparison of the education of participants' children versus control women's. If there was a trend, it seemed to be that the children of the uninvolved women had slightly more chance to complete secondary education than had children of the participants.

In regard to decision-making, a few interesting and significant differences emerge. Uninvolved women are significantly more likely to decide on the use of their own income than participants, although most of the latter do so in conjunction with their husbands. These women with no contact with the factory were also more likely to decide on what their work is and how and where they do it. Both groups said their dwelling was decided on jointly with their husband for the most part, and the majority of both groups of women made the decisions about what the family would eat, although uninvolved women were more likely to make this decision on their own. More of the latter group also decided on their sons' and daughters' education. None of the participant women felt their involvement in the project had any impact on any of these decision-making areas or on the overall decision-making pattern in the family.

Not surprisingly, the participants did not credit the project with having had a major impact on their lives. Eighty-three per cent said it had not had such an influence. Only six per cent said their income had risen because of it, while three per cent said their income had fallen. None of the participants reported receiving any kind of training nor access to credit or a new technology as a result of the project. Seventy-two per cent said it had not affected their self-confidence or views on their future economic prospects. Most of the women reported their husbands liked the higher income they received during their employment in, or trade with, the factory although a few said their husbands had not liked their decreased availability for the family. Eighty per cent of the women suggested the project should be changed in some way although there was no consistent pattern to their suggestions. A majority (62 per cent) also felt the project had had merit, again without a clear designation of why.

In sum, the picture which appears from the data is one where the women who had sold to or worked in the factory had briefly (during its tenure) earned a higher income than they otherwise would have. The project, however, had not had any significant effect on their incomes over time. Nor had it otherwise led to a development of their lives in terms of their status or of the well-being of their family. In other words, the results reflect the absence of any efforts at training, mobilizing or otherwise assuring improvements in the lives of the women who were associated with it. Perhaps had there been a longer-term rise in income resulting from longer or more regular employment or market opportunities, there might have been more of an effect on the women. As it was, most women were glad to get the income but the experience had not changed things for them. Other women who had not had that opportunity had done just as well, or better.

The male participants reflected the same patterns. Because there were so few (ten were interviewed) with only a tiny control (five uninvolved men), they can be dealt with briefly. Most of the uninvolved men were in the high-income group, while most of the participants (80 per cent) were in the middle-income group. Participants were slightly more likely to have land or a house, less likely to own livestock and more likely to own other assets. They were significantly more likely to get income from their assets, where they had them. This was the only statistically significant difference between the groups, but, given the small size of the sample, should not be attributed much importance. Half of the participants had acquired assets during the project period, but none felt this was due to the project. None felt the project had improved their children's educational opportunities or their family's diet. Nine men felt the project had not changed their lives at all, while the other man was not clear what the project had done for him or to him. None felt the project had any impact on their outlook for the future. Most said their wives liked the income they received when the factory was open, and indeed most of the men felt there were some benefits they had gained – in particular, the income they received when the factory was in operation. Thus, the men participants, like their female counterparts, suggest the project had few long-term impacts on them or their lives. From a developmental perspective, then, this venture capital sub-project did unleash capital for investment but only resulted in temporary changes in the income of the targeted beneficiaries.

Expectations and reality

There are several possible perspectives for evaluating the success and the impact of the CTA. Looked at from the vantage point of the principal original goal of the RSSI, the CTA was a worthwhile investment. Capital advanced by the RSSI has been used to leverage substantial local capital and obtain a bank loan probably otherwise unavailable. The combined capital had created an enterprise which had been highly profitable. During the first year, 13 million baht of pickled ginger had been exported. The turnover/fixed assets ratio was 2.88 (4.5 million baht of the original investment were in fixed assets). Fully one hundred per cent of the production of the factory had been exported, all to Japan at good prices. The RSSI had exited as planned, earning a profit to be used for other investments in the local market. Each of the original investors had realized a profit and made their decisions as to the use of this and the equity they held in their own ways. The last original holder, Ms Jitpranee, then sold a majority of her shares (again at a profit). The current holding stage (non production) of CTA is probably less profitable to its current shareholders (Ms Jitpranee and Thai Choo Ros) but the factory and future possible production remains an asset in hand and is still at least 45 per cent a local asset.

The short-term impact on people other than the original investors in the zone was also positive. During the three years that CTA produced pickled ginger, workers were employed in the factory at a wage of 300 baht a day ($11.76). Average agricultural wages in this region are lower than this (see per capita income cited above) so that the workers were highly attracted to this opportunity.[7] Traders and farmers also benefited from the pickled ginger enterprise. Some of the traders were the original investors, and others either worked for one of the original investors or merely sold to the plant. Their function was to purchase from their network of farmers the young ginger for processing. They also distributed better quality seedlings and acted to provide some kind of quality control and quality incentive. For their intermediary role, they took a small cut of the price paid for the ginger plants. In 1990, for example, a trader received seven baht per kilogram of ginger, retained 0.20 baht per kilogram and passed the rest onto the farmers. Some of the traders also grew ginger themselves and sold this to the factory. The farmers themselves gained substantially from the presence of the factory, as young ginger sells for a much higher price than their other crops. The average income per rai of the ginger was about 20 000 baht, whereas a farmer would only realize 400 to 500 baht per rai from growing corn. Moreover, the price paid to the farmers for the ginger by the factory was higher than the prevailing market price.

Once pickled ginger production stopped, farmers either had to find a new buyer (the Japanese company, for example) or revert to their other crops such as corn and green cabbage. Workers had to look for new employment opportunities and continue to grow their household crops. For both, the CTA had given an increased income although only for a short period (three years). Were the objective of the project to establish a small enterprise which would become self-sustaining and enduring enough to provide income for the local populace over a long period, then this project could not be seen as successful. However, establishing a small enterprise as such was not the aim of the project.

The argument for a venture capital project is that with a small investment and few supporting costs (as even initial technical backstopping was paid for by the enterprise itself), local capital could be generated which would benefit the local economy, and from the profits obtained further local capital could be generated, etc. Perhaps Thailand is one of the few countries, certainly the only case considered here thus far, in which this approach would benefit women at least as much as men. In Thailand, the

---

[7] Indeed, as ATI points out, over the three years of pickled ginger processing, the number of hired workers ranged from 300 to 400 and the period of employment ranged from 4 to 6 months a year. Thus, the lower boundary for the total annual wage payments during the period of operations from the enterprise is B7.2 million and the upper boundary is B12 million. Multiplying the average of these two values, B9.6 million, by the three years of operation, the total employment income can be estimated at B28.8 million (or US$1.13 million). Thus the ratio of local wage income generated to the amount of donor finance provided (B1.5 million) is 28.8 to 1. This is a noteworthy achievement that very few development projects ever achieve . . .

economy is open and women already play an active role in almost every sector. But, given this situation, women are full beneficiaries and this has to be seen as a viable approach which can achieve a set of important objectives which will benefit women. Despite the comments of RSSI that their concerns were to provide jobs and generate income and bargaining ability for local farmers (see above), the project's primary objectives were first and foremost to generate local capital with the assumption that, if the enterprise remained local, then others in the local economy would benefit. Judged from that perspective, the project did what it set out to do at low cost. Investors will go on to invest further in the Thai economy[8] and farmers who have profited have ploughed those profits back into their holdings and await the opportunity to respond again should the demand for ginger rise as it did with the CTA production.

From the point of view of the developmental goals underlying this study, however, the Ginger Pickling Project falls far short of any other case examined here. It had a positive set of results in terms of unleashing capital for investment but it had little or no measurable long-term developmental impacts on those who came in contact with it. Certainly it did not, in any sustaining way, alter for the better the lives of either the poor women farmers who traded with it, or the labourers who worked inside the factory.

---

[8] Aside from Ms Jitpranee, Rachitta Na Pattaling notes in her Notebook that two of the other original women investors have left the zone to set up businesses elsewhere in Thailand while Ms Jitpranee is involved in a large and profitable set of businesses in Chiang Rai.

Chapter Seven: Survey Analysis Results in Thailand

1) Sample types (female participant, control) and wage employment in agriculture (works and earns 50% or more of family income, works and earns a fraction of family income)
   Chi sq. = 15, Phi = .866, P. = .0001
2) Sample type (female participant, control) and membership in a co-operative (no, yes)
   Chi sq. = 2.903, Phi = .23, P. = .0884
3) Sample type (female participant, control) and decision on use of income (respondent, her husband, a family member, both she and her husband together)
   Chi sq. = 6.804, Cramer's V = .355, P. = .0784
4) Sample type (female participant, control) and decision on use of work (respondent, her husband, a family member, both she and her husband together)
   Chi sq. = 10.864, Cramer's V =.453, P. = .0125
5) Sample Type (male participant, control) and income from assets (receives none, receives small income from, receives 50% or more of family income from))
   Chi sq = 6.562, Cramer's V = .661, P = .0376

CHAPTER EIGHT
Ghana – Shea Butter Processing

Structural adjustment and development

Ghana is a poor country with a population of 14.9 million living in a land area of 239 km². The average annual per capita income is only $390, higher, of course, in the capital (Accra) and other urban areas, substantially lower in non-cocoa growing agricultural areas. Seventy-five per cent of the school age population is in primary school, 39 per cent in secondary school and only 2 per cent in colleges or universities. Forty per cent of the adult population is illiterate. Ghana's overall rate of growth in per capita income from 1965 to 1990 was a negative 1.4 per cent, but this disguises some recent more positive trends. From 1965 to 1980, the GDP grew at an average annual rate of only 1.3 per cent, but, from 1980 to 1990, a substantial improvement was realized when the rate for these years rose to three per cent per annum. Ghana, however, suffers from severe inflation problems (although not as severe as Peru). The average annual rate of inflation between 1980 and 1990 was 42.5 per cent, double the rate that existed in the fifteen years before. Ghana has a primarily agricultural economy with 48 per cent of its GDP coming from that sector, only 16 per cent from industry and 37 per cent from the service sector. In fact, the percentage contribution of the agricultural sector has grown since 1965 and the share of both the industrial and service sectors has decreased (World Bank, 1992: 218, 220, 222, 274).

This somewhat curious decline in the relative contribution of industry and the service sector to the GDP needs explanation. Ghana had a thriving cocoa production on which its export economy was based since its independence in 1957. By the mid-1960s, however, Ghana was in severe economic trouble. The fall in world cocoa prices and problems in production were combined with an over-extended state sector, and over-valued exchange rates. Thus, although all of Africa south of the Sahara had an annual 3.8 per cent growth average for the years 1960 to 1970, the economy of Ghana contracted during this period. The first President, Kwame Nkrumah, had emphasized the role of the state in building the economy, and the importance of investing in industrialization and building the urban infrastructure as a means to this end by using the profits from cocoa production controlled by state marketing boards. Nkrumah was overthrown in 1966. Faced with weak, non-productive state enterprises and a stagnant private sector, the regimes which succeeded Nkrumah were, by and large, ready to reduce government control and encourage private investment. A further blow to the Ghanaian economy, however, occurred with the world rise in the price of oil in 1979. Ghana, like most other countries in Africa south of the Sahara, was badly hurt by this development. Ghana found itself with a huge debt and declining resources to meet its payments. Indeed, in the late 1980s, Ghana's total external debt service payments were more than 50 per cent of its exports (Chazan et al, 1992: 305–10).

Ghana was one of the many African countries which approached the World Bank and the IMF to negotiate a structural adjustment programme in order to reschedule its debts and obtain additional financial assistance. The programme, agreed on in 1983, included six principal elements:

1) Devaluation of the currency: the cedi was devalued 6000 per cent in nominal terms (actually 90 per cent in real terms) from 1982 to 1987;
2) Reduction in tariffs and quantitative restrictions on imports to Ghana;
3) Decreasing government expenditures to reduce the budget deficit, including ending agricultural subsidies;
4) Divestiture of public corporations and general dismantling of state-owned enterprises;
5) Agricultural reforms, including increased producer prices (without a marketing board to skim off the profits) and reduction of government control over agricultural operations even including the provisions of inputs; and
6) Restructuring of the public sector through a variety of institutional reforms (Chazan et al, 1992: 108–9).

As a result of agreeing to these reforms, Ghana benefited from five IMF programmes and more than 20 loans from the World Bank from 1983 to 1989. It became the showcase of structural adjustment, widely touted as proof that this policy could work. Indeed, between 1983 and 1989, domestic production in Ghana grew six per cent per year. The rate of investment also doubled in this period and the budget deficit was cut by two-thirds while exports increased dramatically. In this context of declining state subsidies, the agricultural sector regained some of its earlier predominance.

However, some analysts do not find the Ghanaian results quite as positive as some reports from the World Bank have claimed. Fantu Cheru, for example, points out that a close look at the Ghanaian economy shows that structural adjustment did not achieve as strong economic gains as had been claimed. A small minority of cocoa farmers (32 per cent of those growing cocoa) did increase their income substantially, but the rest did not, and the income of the non-cocoa agriculturalists stagnated. Food self-sufficiency in the country actually declined. In addition, the world market price of cocoa has been dropping and Ghana's terms of trade have seriously deteriorated. To make up for declining foreign exchange from cocoa, the timber industry has received support from the World Bank, with the result that forest areas in the country have been drastically reduced. At the present rate of depredation, it has been estimated that Ghana will be fully stripped of trees by the year 2000 (Cheru, 1992: 506–7).

Devaluation of the cedi was specifically indicated as a cause of suffering for most Ghanaians. For example, poor farmers were unable to purchase fertilizer and other inputs, the prices of which had soared as a result of the lower value of the cedi (Cheru, 1992: 507). Nor was it only those in the agricultural sector who suffered. In the industrial sector, which had expanded in the wake of structural adjustment, complaints were registered that the expansion was limited because of devaluation. Business owners complained that they could not afford the cost of inputs and spare parts although dropping tariff barriers had supposedly made these more available than before (Osei *et al*, 1993: 53–8).

In fact, the Ghanaian debt did increase from US$1.3 billion in 1983 to 3.1 billion by 1990. The fragile new industrial sector saw little expansion, and the burgeoning cocoa production faced major competition from increased cocoa production in other countries, thus driving down the world price for this commodity and leading to a 'crash' of the cocoa market in 1988, as mentioned by Cheru above. Domestic investment did not grow as expected and even international investment was limited (Chazan *et al*, 1992: 313).

In one way, Ghana differed from many other countries experiencing structural readjustment because it specifically attempted to deal with the social consequences of these reforms. Decreasing government expenditures means that government programmes such as education, social welfare and health are cut back. Permitting the price of food to rise according to the market means a substantial increase in the cost of living for urban dwellers. In these circumstances, the poorer sectors of the population were bound to suffer and the country faced the risk of increased disease, poverty, and the slowing down of the spread of mass education. In Ghana, a Programme of Action to Mitigate the Social Costs of Adjustment (PAMSCAD) was adopted. One part of its programme related to small and medium-scale enterprises which the current government of Ghana, under the direction of President Jerry Rawlings, has attempted to encourage and support. Probably the most important way in which this concern is implemented is through changes in the structure of tariffs, interest rate policies, ending import licence requirements and controlling the price of inputs. Small entrepreneurs had been disfavoured and kept out of the product market. With the new lowering of barriers, small entrepreneurs were free to negotiate, to buy the parts and supplies needed and to operate with the changing levels of market demand.

Even before the structural adjustment programme, the governments following Nkrumah had attempted to support the growth of small enterprises. In 1970, the office of Business Promotion, later renamed the Ghana Enterprise Development Commission (GEDC), was established to help the development of small-scale industry with both technical and financial support. After the structural adjustment programme had been adopted, these efforts to reach small entrepreneurs were intensified. The National Board for Small-Scale Industries (NBSSI) was set up within the Ministry of Industries, Science and Technology. In 1987, the latter established the Ghana Appropriate Technology Industrial Service (GRATIS) to 'help upgrade small-scale industrial activities by means of a transfer of appropriate technologies to small-scale and informal sector industries from the grassroots levels'. Regional divisions, called Intermediate Technology Transfer Units (ITTU), were set up in six regions: Tema, Tamale, Cape Coast, Ho, Sunyani and Kumasi. These ITTUs were supposed to develop the engineering capabilities of small-scale industries in the fields of vehicle and related repair trades, and to support non-engineering industries by the manufacture and repair of machinery, plants and equipment. In addition, in 1988, the Bank of Ghana received a $28 million credit from the International Development Association of the World Bank to establish a Fund for Small and Medium Enterprises Development (FUSMED). PAMSCAD also established a revolving credit fund of two million dollars to assist small-scale enterprises (Osei et al, 1993: 53–8).

Not all the impacts of new government policies have been favourable for Ghanaian small enterprises. For one thing, opening the economy by reducing tariffs and quotas has meant competition from well-established foreign firms which some fragile new enterprises can not tolerate. Secondly, the loss in value of the currency meant that the cost of inputs and parts became increasingly expensive in the informal as well as the formal sector. Thirdly, the lack of credit and financial backing continued to be a major factor preventing new enterprises from being established and old ones from expanding. Overall, however, the more free and more supportive environment has had a positive impact on small-scale enterprises. In a recent survey of small enterprises in Ghana, forty-two per cent had been established after structural adjustment reforms had been adopted. Moreover, the output of most small enterprises increased in this period. Fifty-three per cent of the survey indicated such an increase, most of which expanded in the range of a zero to ten per cent increase, while sixteen per cent, mostly larger enterprises, expanded as much as fifty to one hundred per cent (Osei et al, 1993: 58–70). As a result, the climate for small and medium-scale enterprises is generally favourable in Ghana at the present time, which is an important consideration in evaluating the Shea Butter Project.

Women in the marketplace

The situation of women in Ghana bears a certain resemblance to that of women in Thailand, in terms of the relative independence and the economic involvement of women in both places. There are, of course, major differences as well. Among the many ethnic groups of Ghana, some are matrilineal and others patrilineal. This variation has often been linked to increased economic and political power for women in matrilineal situations, as compared to the lesser power of women in patrilineal family structures. However, Ghanaian women in both types of situation have a relatively 'high degree of socioeconomic independence' except in the Muslim areas in the north. Transfer of property and money is still largely through blood relations rather than marriage, and children have strong obligations not only to their mother and father but to their extended lineage. Marriages are unstable, with as many as fifty per cent or more ending in divorce. In these circumstances, women are expected to be major sources of income for the family. Even when married, they are expected to provide a portion of the food, clothing and educational costs for the family. In rural areas, women fulfill these obligations through their work on the family fields and on their own plots. They sell their surplus produce in local markets and sometimes to middlemen traders. In urban areas, women are often involved in petty trade, sometimes of products prepared at home or with other women, such as food, textiles and other commodities (Steel and Campbell, 1982:237–8; Vellenga, 1986: 62–77).

The introduction of various incentives to the modernization and development of the economy have not all served the interests of women. For one thing, the increasing registration of lands in the names of private individuals lessens the importance of the extended family. As land is most often registered to the male, women lose access and economic power. In addition, the process of development causes strains in the traditional balance between male and female labour and responsibilities. This has resulted in tension and antagonism which has been reflected in government policies to undermine the economic status of women or to prevent any expansion of their economic roles. Thus, women entrepreneurs have been attacked both literally and in the press. For example, during his first term in office in 1979, the current Head of State sent soldiers to raze the Makola No. 1 Market, the centre of retail trade in Accra and dominated by women traders, with the comment, 'That will teach Ghanaian women not to be wicked.' Their 'wickedness' seems to have been their involvement and success in trade. In addition, governments before structural adjustment imposed price controls and taxes, offered no government infrastructural support to markets, and made direct attacks in the newspapers on women's enterprises (as greedy, immoral or acting contrary to explicit laws) to discredit women entrepreneurs and discourage the growth of their enterprises (Robertson, 1984: 238–47; Vellenga, 1986: 76; Roberts, 1987: 48–67).

In the current era, the Government of Ghana, strongly encouraged by the IMF, the World Bank and other donors on whom it depends, is making a more concentrated effort to help women entrepreneurs. A recent survey of small enterprises found only 26 per cent of the entrepreneurs were women, suggesting to the analysts that the finding 'points to the inadequate attention given to encouraging women to commence in business' (Osei *et al*, 1993: 63). A different sampling process, one which looked at a random sample of enterprises rather than a stratified one, might have found more women in the small and more numerous food production businesses which would have increased their proportion of the whole. However, from the point of view of this study, this conclusion is probably warranted. Women have not been encouraged or supported in ways specific to their education and family situations. They have been actively discouraged from being entrepreneurs and this has affected their move into this sector. The current project is one effort to reverse this trend.

A final comparative look at the role of Ghanaian women, shows that forty-nine per cent of the women of Ghana are illiterate, but this is not widely different from their male peers. Sixty-seven per cent of the relevant-aged girls go to primary school, thirty per cent to secondary school and two per cent to college or the university, only slightly less than the percentage of boys who do at the same ages (World Bank, 1992: 218, 274, 280). As the industrial workforce has expanded, women have begun to enter it. They remain in lower status and less well-paid jobs, but they are beginning to be hired in all sectors of employment. In 1990, women were ten per cent of all administrative and managerial workers. Almost three times as many women as men held jobs as clerical, sales or staff workers, and eighty per cent of the agricultural workforce was female (UNIFEM, 1991: 105). Most adult Ghanaian women expect to be actively involved in work outside their homes. They need to earn money on their own account and, as with their male counterparts, the current climate, with only a slowly expanding industrial and service wage sector, makes small and medium-scale enterprises an attractive and viable choice.

The Shea Butter Processing Project in Ghana

One traditional economic activity throughout the north of Ghana (and throughout the Sahel where the shea nut tree grows) is the production of shea butter for consumption and sale of shea butter. Traditionally, shea trees grew wild in the savanna area of Western and Central Africa. Women collect their nuts and process them, usually in village women's groups as the labour involved is lengthy and arduous. Each woman would have her share of the product for her own use and disposal once the process was finished. The main product is shea butter, which is used as the primary local source of cooking fat. Estimates suggest that the average daily consumption of shea butter in Mali and Northern Ghana is 20 to 30 grams per person. Shea butter is also used as a pomade for hair and for skin diseases such as boils, for cuts, for the manufacture of soap and as fuel for lamps (Sjostrom, n.d.: 1).

In the nine main producing countries, about 630 000 tons of shea kernels are produced each year. About four-fifths of this production is not used locally but exported to industrialized countries where the fat is extracted and used in place of cocoa butter and for cosmetics and pharmaceuticals. The process of extraction in industrialized countries is estimated to be 85 per cent efficient. In contrast, although the Dagombo women of northern Ghana (where this project is located) are said to have the most efficient traditional extraction process, their extraction efficiency is estimated to be only between 30 and 35 per cent. Their traditional process of extraction involves ten steps:

1) Women and children collect the mature nuts which fall to the ground.
2) The nuts are brought to the family compound and boiled for two to three hours to dissolve the fleshy part of the fruit and kill off harmful micro-organisms and enzymes.
3) The boiled nuts are dried in the sun for two to five days.
4) The nuts are deshelled manually in a mortar or on a flat surface with a wooden pestle. Then the shells and kernels are separated.
5) The kernels are dried in the sun to allow for storage and then kept in jute bags (for as long as two to three years if necessary).
6) The kernels are crushed to a granulate of more or less uniform size in a wooden mortar, usually by two women each with a pestle.
7) The kernels are roasted in big round pots placed on three stones over a wood fire. They are constantly stirred to prevent sticking. The roasted mass is placed on the floor for cooling.
8) The roasted kernels are ground to paste, traditionally by a group of fifteen village women working in relays of three. The paste is made very smooth by grinding small amounts between two stones.
9) The cooled paste is placed in steel basins, and then kneaded or churned. A group of six women stir the mass while adding water to an equal amount. When this process is finished, about five litres of water are added while the contents are stirred and a thick, spongy whitish layer, the shea butter, rises to the surface.
10) The washed 'cream' is heated, the scum discarded and the remaining oily residue removed. The clarified oil is kept in clean enamel basins and left to cool. It is stirred regularly until it begins to be semi-solid. Then it is transferred to a calabash and shaped to a round white or yellowish lump, its final form, which is brought to the market to be sold (Sjostrom, n.d.: 4–14).

The Shea Butter Project was designed and adopted by GRATIS in 1989 through its Tamale ITTU to promote this traditional women's enterprise. By disseminating a set of equipment for small-scale production of the butter, they hoped to reduce the labour required, increase the rate and amount of production, improve the value of the product, increase the income of the women's groups involved, and increase the local food supply. The initiator of this project was the National Council on Women and Development (NCWD) which was already working with women in this zone. The NCWD asked the ITTU to develop a shea butter project for the women. The Technology Consultancy Centre and the Department of Biochemistry at the University of Science and Technology at Kumasi had done research on shea butter processing in the 1970s. In 1987, GRATIS consulted the TCC, which suggested three areas of intervention. First, the crushing and milling could be done mechanically with a corn milling machine. Second, the most critical part of the process was the kneading (or churning) and a machine should be developed to help with this phase and, third, the roasting of the kernels and boiling of the cream should be altered in order to decrease the use of woodfuel and exposure of women to the fire (Donkor, 1991: 1). The TCC asked one of its client organizations, SIS Engineering Ltd, to develop a kneader. This kneader was pilot tested in the village of Bunglung in 1988.

Bunglung, and the succeeding villages in the project, were chosen by the NCWD for the project because there was a women's group in existence which traditionally produced shea butter. At least 70 per cent of the village women were to be involved in the process of shea butter production. There had to be abundant shea trees nearby (60 per cent of the trees in the village area had to be shea). After the preliminary pilot test, fifteen new sets of machines were ordered from SIS and installed in various villages supported by different grants from international organizations through the medium of several different Ghanaian agencies. These sets of machines had not been thoroughly tested and their introduction was not uniformly monitored. Therefore a new test phase was undertaken in 1990 in four villages supervised by the ITTU. The villages, chosen by the criteria outlined above, had different relationships to the project (see below).

The project originally included five-day seminars to train the women in the use of the equipment (and other courses of differing lengths to train personnel in the ITTU, and other Ghanaian agencies, in manufacturing the equipment and backstopping its usage by the women's groups). In the later stage, a more extensive two-week training seminar was adopted for the village women with demonstrations, and training in the use of, the equipment. Four hundred and forty-three women from village women's groups have been trained in these seminars.

Women's groups in the chosen villages were given a loan for, or had to purchase outright, the production set. Each set included a corn mill, a crusher, a kneader, and an 8-HP diesel engine for a total of C1 160 000. Added to this were the costs of installation and training (C307 000) and the costs of building a shed for the equipment (C1 600 000) for a total of C3 067 000 or $8713. The running expenses, including fuel, cooking oil, the ground plate and the costs of the mill operator, came to C1 326 000 or $3767 per year. Maintenance costs were C10 600 or $30 for the same period. These costs were to be offset by service fees the women's groups could charge to other women who were without this equipment for processing their shea nuts, at an estimated yearly fee of C2 925 000 or $8309. This sum was enough for women's groups to pay not only running costs and maintenance, but also interest and principal over a period of several years where they had borrowed the money for the initial costs. Moreover, the project estimated a clear and very large gain in net income to the individual women participating in this project compared to those who still processed shea nuts in the traditional fashion (see Annex 3 for details).

Because many different agencies were involved in this project at different stages, it is impossible to fully report on the costs of establishing and implementing it (which is still not completed). This is a serious problem in evaluating this project compared to other projects. Our approximation of the project costs (including materials, costs of producing sets, overhead, travel and salaries or fees) is $150 000.[1]

ITDG supports GRATIS through grants and technical assistance. In the case of the shea butter project, ITDG provided consultancy support and specifically helped in the negotiation for the export contract with the English purchaser, Body Shop.

The project in the field

The Shea Butter Project began in 1989 and was, in 1993, still in process in its second phase or project extension. There were eleven villages directly associated with the project (but with varying types and degrees of support), and 423 women who benefited directly from the project.

In the initial stage of the project, in 1989, both the SIS kneading machine and a different version, which GATE/GTZ had tested in Mali, were introduced in Bunglung in 1989. The Mali alternative was rejected by the women because the rate of extraction was low and the final product was in the form of brown oil rather than the traditional 'cream' they were used to. Thus, the SIS machines were adopted for use in the rest of the project. Thereafter, fifteen machines were purchased and installed in various villages by UNDP, the Canada Fund and others.

In the second stage of the project, which was funded by GATE/GTZ, four machines were to be tested under closer supervision. This stage, originally scheduled in 1990, was delayed by a lag in the transfer of funds and by the inability of the village women to find the one million cedi necessary

---

[1] Without information on the costs of the project manager, technicians' fees etc., we can not know what the total costs of this project were. $150 000 is only a guesstimate.

for the construction of the shed where the machinery would be housed. The shed was to be constructed by the Department of Rural Housing and Cottage Industries using low-cost techniques, but the materials for the roof, cement, etc. were nonetheless costly for the villagers. They sought funds from the PAMSCAD programme mentioned above, but PAMSCAD was able to fund only one village and then, only to the amount of 400 000 cedi (ITDG, n.d.: 1).

Ultimately, four sets of machines and a diesel engine were installed. Two of these were supplied by SIS Engineering. SIS also trained Tamale ITTU staff in how to make the equipment and they produced the other two machines, one of which was used as a project demonstration, and the other by a village women's group in the normal manner.

The project had a different impact in each village depending on the local circumstances and the donor's programme (the amount of training, whether or not loans were made, exactly what machines were introduced, how much technical back-up was available, etc.). One example illustrates what could happen. In Bunglung a women's group was already active. This group had been started in its present form in 1983 after the Volta Dam had been constructed. The group had twenty-two members and was headed by a president, a treasurer, a secretary and an animator. The treasurer, the village schoolteacher, was a man; the women were illiterate. At that time, women had two principal common activities: growing vegetables during the dry season and processing shea nuts during the rainy season. NCWD had been providing them with extension support. By 1984, they had ceased the vegetable production because there was not sufficient water for the vegetables. In the traditional fashion, they divided up the shea processing work. The final product was divided among the women who each sold their own.

The women were put in contact with the United Nations Development Fund by NCWD. The UNDP, through their Technologies for Rural Women Program (with Dutch funding of C1 752 235 or $4978), funded the project for this group. UNDP purchased the first set of machines and trained seven women's group members in equipment maintenance and twelve others in small business management. C200 000 ($568) was transferred to a bank account to meet the group's working capital needs. Because of problems with the first set of machines, a second set was purchased by UNDP under a new programme: Technology, Capitalization and Training for Rural Business Women (which was taking place in 22 Ghanaian villages). An additional C200 000 was provided to the women for further training. The new set of machines was introduced to Bunglung and a mill attendant was hired (as had been for the old set) to run and maintain the mill. Now, during a normal week, the women devote three days to individual processing and three days to community production. The nuts are bought on the market for the group as a whole, crushed, and then divided among the women for home processing. Three or four maxi bags are processed collectively and the final products sold to the market. The common profit is put into a savings account. An additional profit comes from charging other women for milling. The savings are divided between maintenance and repairs (C200 000) and a reserve for buying shea nuts (C60 000).

The new set of machines has cut the processing time for the Bunglung women in half, but they still face problems in bringing the water and wood needed for the processing and in transporting raw materials and the final products to the market. In addition, the kneading machine was broken in spring 1993. The cooling component did not seem appropriate to the women who wished to replace it with a water cooling system. The latter, however, would cost between C600 000 and C1 500 000, which they can not afford. They have already had to pay for the repair of the mill stones and the replacement of other component parts.

The experience at Bunglung indicates some of the difficulties which women participants have faced. Indeed, three persistent problems have plagued the project in most villages. One of these is continuing problems with the technology. A study done on the first machines released by SIS indicated that, out of four kneaders introduced, three had been used satisfactorily for some time but none was in continuous use, nor had they been fully accepted by the women. The machine at Bunglung illustrated the problems. At first, the speed of rotation was too high and this had to be adjusted down to 50–60 rpm. A hand crank had to be installed to allow an even slower rate at the end of the process. Women asked for easier access to the top of the reactor where the butter was forming so they could monitor the process more closely. Temperature was also a problem. The women were unable to maintain the temperature in the kneader at 30–32°C, the temperature required for creaming. The solution to this problem is the replacement of air coolers with water coolers. Finally, women were unable to make small repairs to the machinery when it was in operation, which would make simple breakdowns less likely. By 1993, six new machine sets had been installed and three more were being considered. Technical problems still exist and many machines are broken through improper, or lack of, maintenance. Furthermore, the English diesel machines appear to be too expensive for the villagers while the Indian ones seem to break down too easily.

A final technology problem is that women still have to fetch wood, haul water, boil, roast and knead the nuts. A boiler to simplify the process had been envisaged, but it was not accepted by the local women who continue to roast and boil manually at home.

The second major problem in the project is product marketing. When traditional production was common, less was produced and all could be sold locally. Since the machines have been introduced in several villages, the amount produced is much greater and has outstripped the demand of the local market. Women often come back from the market with their shea butter unsold. The high price of transport discourages traders from buying in the villages. One positive event occurred as a result of a BBC ITDG film on the Shea Butter Project: the owner of a chain of cosmetic product shops called the Body Shop (Anita Roddick) decided to place orders for Ghanaian shea butter – 12 to 16 tons of shea butter a year. A local company, Wonoo Venture, would be responsible for the export of the butter. This outlet, however, had not yet been activated in 1993; Ghanaian law requires a six-month period before an export licence can be obtained. Nor will this export outlet alone be enough to absorb all the product of the now highly producing villages.

The third problem is the village women's groups' lack of capital. This has made it difficult, and in some cases impossible, to meet the needs for the construction of the shed for milling for women in many villages, and it has hampered women from buying the shea nuts during the season when they are cheapest and storing them for the off-season. The costs of proper storage facilities are too high. Lack of capital is also one reason why machines are not repaired or maintained properly.

Many faces, many experiences

Forty questionnaires, divided among five villages, were administered to women who had participated in the project (three villages had ten respondents and two had five). A control sample of thirty women from three nearby villages completed the women's sample while fourteen husbands of women in projects also received questionnaires. All of the women reside in the zone where village women generally process shea butter. A majority of both groups were aged between 21 and 40 years old. Few of the women were heads of their households although slightly more of the non-participant women were. Most women were married but the participant women were more likely to be widows than the other women. All but three of the women were Muslim. Most of both groups had a co-wife. Most of the participants, however, were not the first wife, while most of the other women were. However, the difference between the samples was not statistically significant. A plurality of both groups had four to six children while the next largest group had seven to ten children. Only two women had no children.

More than 97 per cent of both groups of women had never had any schooling. The participants, however, were significantly more likely to have had some literacy training; more than half of them had done so. More than eighty-six per cent of both groups said their husbands were farmers, while there was also a scattering of agricultural labourers, teachers, public employees and one listed as unemployed. The husbands had had no education, although three husbands from the non-project sample had had literacy training and one participant's husband had completed part of secondary schooling (he was the teacher).

The Ghanaian women clearly did not think of themselves as having an occupation. The participants all said they were housewives and the other women could not identify an occupation. The women were equally unclear about their principal economic activities, although in this area of Ghana they are involved in raising rain-fed crops, vegetables (when water is available), chickens and other poultry, sometimes goats and other small ruminants, and other economic activities such as making charcoal and, of course, processing shea nuts. None of the non-project women felt their economic activities had changed during the years of the project but most of the participants did think this had happened. They (83 per cent) felt they had changed what they did and now received a larger income. Most of them (83 per cent) believed the project had been the reason for the change.

Most of the participants had not been in the project for very long, only one to two years for most. They were divided as to why they joined it. Some women (26 per cent) simply said that they hoped to get a greater income. Some identified the new technology used as the reason for joining, while others said the project had promised to get credit, training and access to the new technology for their co-operative. In fact, all of the participants were members of a co-operative (this was a pre-condition for being allowed in the project) while few of the other women were. Many of the latter did belong to some kind of economic group such as a credit association, but they were still significantly less likely than the participants to belong to any kind of economic group.

Interestingly, the families of the participant women were not better off than the families not associated with the project – indeed it was quite the reverse. Most of both groups were living in below-subsistence poverty but the non-project women were much more likely to be in the middle-income bracket than were the participants (this difference was only significant at 10 per cent) despite the much higher profits from processing the shea nuts with the new technology received by the participants.[2]

More of the participants than the other women said their husbands had no land or house, but this difference was not significant. On the other hand, while most of both groups said their husbands owned livestock, significantly more of the participant women made this claim. Most of both groups said their husbands received some kind of income from their assets. In regard to the women's own possessions, there are stronger differences between the groups but they are contradictory. Significantly more participants claim to

---

[2] The annual income range is established at: high = 2 100 000; middle = 480 000 to 2 099 999; low = 205 000 to 479 999; below subsistence at less than 205 000.

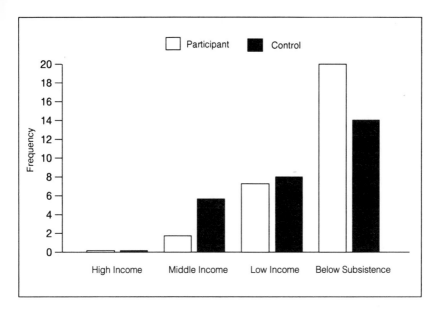

Figure 8.1 *Income levels*

own land or a house while significantly more of the other women say they have other assets, in this case primarily jewellery. Three-quarters of both groups say they have no livestock of their own. More than ninety per cent of both groups do not get an income from whatever assets they own. We are left, then, with an unclear profile where the participant women seem poorer than the other women, yet they are more likely to have land than are the non-project women. Distinctly more of the latter, moreover, have jewellery, the traditional wealth of women.

None of the participants said they had acquired any assets during the project period and most (78 per cent) said the project had not affected their ability to acquire assets. There was a strong sentiment, however, that the project had provided them with a greater income. Although twenty-one per cent of the women said it had not affected their income, the others either said the project had provided them with income or somehow improved their situation in regard to their assets.

Food is the primary item that both groups of women in these villages purchase with any income they receive. Twenty-two per cent of the participants, however, say they have changed their expenditures over the last few years (which is a significant difference from the other women). These women say that before they had had to spend the money on clothes for their children or themselves and do not now. Asked directly if the project affected their expenditure pattern, most participants did not respond. Of those who did, most said it had not, but a few of them indicated some effects; two said it had provided them with more money to spend, another two said it had caused them to incur debts!

Although the financial profile of the two groups is not sharply different, some aspects of the lifestyles of the two groups appear to be quite distinct. This is not true for diet or children's education. Neither group had access to luxury foods, such as regular beef or mutton (although a majority (58 per cent) of the participants said the project had helped them acquire better food), and most women of both groups had children in, or who had completed, some part of elementary school. In regard to decision-making and time use, however, the two groups diverge widely. Participant women said that they now spent more time with their children and working on their household chores. They also said they had more time for their own education and self-development *and* more time for rest and leisure activities. In addition, they spent more time on gathering fuel and finding water. All these patterns of time use were significantly different from the other women who had generally made no changes. Virtually all of the participants credited the project with changing their use of time, with about a third saying clearly that it had provided them with more time for themselves and for their family than they had had.

The profile of the participants is even more dramatically different when their status in family decisions is considered. Sixty per cent claim to decide how their income is used and on whether they work outside the home and what they do. Decisions on where the family lives remain a male prerogative for both groups as does the decision on what the family eats, although participants are slightly more likely to say they are consulted on this or make the decision themselves. Their husbands also decide for both groups on how much education both sons and daughters will receive, although participants are somewhat more likely to participate in deciding their daughters' schooling. Despite these differences, the participant women do not generally credit the project with having given them their current role in family decisions. Most say it had little or no impact on their power in the family and most say they have not changed their role in the family during the project period in regard to any decisions except use of their income. Curiously, there is a significant difference between the participants and other women on this one dimension, and the difference is negative from the developmental point of view. Thirty-eight per cent of the participants say they had decided on the use of their income before, but now their husbands have taken over this decision. This, in fact, may reflect the greater importance to the family of their increased income from shea butter. Ninety-four per cent, however, really do not think the project had any impact on their decision-making power.

Ghanaian women are appreciative of what the project has offered them even though they do not think it has changed their family status. Over ninety per cent of them feel the project has provided them with something positive.

Many feel the project gave them a higher income, or simply a better life, or some combination of the two. Most of the women acknowledge that the project trained them in the use of a new technology. Although most do not think it affected their family position or their family life overall, they do think the project made them feel better about themselves and their economic future. Nor do the men think their wives' participation has affected their family. Sixty-four per cent say it had no impact, and only thirty eight per cent said the additional income it provided affected the family circumstances. The women, however, do not seem to be aware – or do not report – this lack of appreciation on the part of their spouses. Virtually all of them say that their husbands like the project for the money it brings to the family.

A closer look at individual responses gives some depth to these statistical results. In Dalung, for example, the women were older than most women surveyed, almost all being over 45. More than half of these women were widows. Despite their age, their attitude to the shea project was openly positive and enthusiastic. They knew what it had done for them. Most said the introduction of the new technology had both decreased processing time and increased production. They commented on their reduced workload. One said that the drudgery of her work had been reduced and now she could do other things at home that she wanted to do. A second, echoing her, said she had more time for herself. Another said that, because of the project, she had increased income to provide more education for her children. Two others said that now they could be certain that their grandchildren will be cared for in the future.

What was also interesting about the Dalung group was their commitment to 31st December, the women's group assisting the village women. Many of them said that working together with other women was very good. They credited 31st December with helping them learn not only about shea butter production, but also farming and even literacy training. They placed a lot of importance on their membership and felt it had done a lot for them.

Gbabshie in Tamale has a much larger group of women involved, many of whom are young. (In our sample all but two were under forty.) 31st December also supported this village. Here the enthusiasm was equally evident. They commented on their lessened workload and on the resultant increase in time for other activities. Several of them said that the income which they now received allowed them to pay for their children's education, which they could not do before. They mentioned the school fees and the cost of uniforms which they could not previously afford. One said she could now contribute to the family expenses, which she could not before, while another spelled out her new contributions saying she could now add fish and salt to the family meals and could use more shea butter in preparing them. One woman summed up the feelings of this group saying 'I can raise my voice in public and the mill has brought some civilization into the village. If well kept, it will be beneficial to the next generation . . .'

Similar comments were made by women in Yemo, Busunu and Nakpayili. The only difference in the responses of the women in these towns was that they did not emphasize as much the importance of the group of women working together or of other women's group activities. This may reflect the fact that they had not been mobilized by 31 December. In regard to the project itself, however, their reactions were very similar.

The final review of the results of this survey suggest certain clear points. The participant women show signs of having gone through a process of development which places them in a position of higher family authority than their peers and makes them liable to have higher levels of education, or at least literacy training. This seems to have been due to their membership in an organized women's group which was a prerequisite for being in the project. Thus, despite other reports (see evaluation below) which say that Ghanaian women do not want to sell their product as a group (individual ownership and sales being the traditional pattern), group membership and group activities are accepted and seen as important. The value of this experience was particularly underlined by those who had been most actively mobilized by 31 December, which apparently places great emphasis on women taking charge of various aspects of their lives as a group. Most of the non-project women had not had this experience, even in the more limited form of group mobilization which the participants in villages without 31 December had received. Thus, it seems probable that women's group membership and its associated activities, rather than the short exposure to the shea butter enterprise in itself, is the reason for these developmental changes, although the shea project reinforces and extends this experience.

But certain changes were not in the direction of development as defined. For example, changes in decision-making that are acknowledged by the participants as a result of the project occur where husbands are reasserting their authority over their wives' income. This may be because the wife's income is now more important to the family. The loss of independence is not, by the values adopted in this study, a desirable outcome. Yet the women participants did not complain. They appeared to feel their greater income was more of an asset than any interference by their husband in their decisions about its use was a hindrance. They had acquired importance and choice as result of being able to contribute more. In contrast, the project certainly must be credited with a strong, positive (by our definition) change in time use. The participants have more time for themselves and their families, although they do now have to spend more time gathering fuel and water – all of which most of them credit to the project. Clearly, the technology did reduce their labour time substantially.

To date, the participants and their families have not emerged as significantly financially better off, although most believe the project has increased their income and changed their economic activities. The participants are more likely to have co-wives and also not to be the first wife than the control women, and the type of complex family structure which this implies may prevent a perception of overall income impacts on the family. Furthermore, and this must be stressed, the women had only a few years of project experience. Their groups were still involved in repaying the loan they received. There had not yet been time for an accumulated income increase which might be registered in larger family income changes. As the project continues, assuming a market for the shea butter is assured, it may have more of an influence on family income and well-being.

Finally, the influence of the Shea Butternut Project is interactive with the pre-existing and continuing influence of agency/group efforts at mobilizing village women to form women's groups and supporting the wide range of activities which these groups undertake. In the Ghanaian case, the importance of this effort to the women who have experienced it is clear. Our results suggest a strong endorsement of the criteria adopted by TCC in its project design, as group membership is a strong factor in the successful outcome of the work.

The reality of the Shea Butter Project

The Shea Butter Project is an interesting case of collaboration between Ghanaian agencies and various international donors. Several Ghanaian agencies were responsible for the development of the project. UST was engaged in the initial research on appropriate technology and feasibility. It withdrew after SIS had been engaged. ITTU took on the role of disseminating and later constructing the machinery as well as helping with training and technical backstopping. The various international agencies together with Ghanaian agencies which work with women (such as NCWD and DSR) are responsible for working with the women: providing funds or credit for purchasing the machines, training women's group leaders and assisting women in organization, leadership and the resolution of various socioeconomic problems. There is, however, no one agreed-upon set of policies for the latter activities. Some village women's groups receive the machines as a grant, others have to pay back a loan for the purchase of the equipment, at least in part. Village groups have had varying experiences in their efforts to build the necessary sheds for the equipment. The amount of training and back-up also varies among the different donor financed shea butter projects.

On one level, the lack of co-ordination among the various groups working with the village women can and should be criticized. No one has taken an oversight or co-ordination role which would allow the experiences of the various strategies to be evaluated and modified uniformly. Additionally, there are still technical problems which have not been fully worked out.

These are related to an incomplete acceptance by some of the village women of the whole concept of group processing and sale. Most women still prefer to process individually although the machine was designed for community use. This means the machines are under-used. Two villages have managed to develop a pattern of sharing time between collective production, the outputs of which are used for maintenance, and individual production for the rest of the week. Most villages have not worked out this problem. Nor have the projects resulted in a high demand by other women's groups to obtain these machines. Women do not perceive the profits because these have been placed in savings by the participating groups and individuals, rather than resulting in a visible change in the standard of living. This, however, may change as the project continues and the participating women pay off debts and gain confidence in their machines and the processing results.

The marketing question has another interesting dimension. The Body Shop is willing to purchase the locally processed shea butter. Other clients in the UK, North America and Switzerland are also interested in the product. If they begin to purchase the butter, the prices they would pay would be higher than what the women could receive on the local market; the local prices vary with the season in any case. In the present situation, increased production of shea butter has meant that some women have not been able to sell their product immediately, but sale overseas might change that scenario. Given that there are a limited number of trees, it is conceivable that (if the overseas demand were high enough) all the shea butter could be sold abroad as these prices would be much higher and the returns better for the producers. In such a case, shea butter would not be available on the local market and it is, as discussed above, a major local commodity used for food, medicine and cosmetics. The success of this project, then, might result in hardship for local women generally.

One solution proposed by the Body Shop is to buy at the local market price and donate the difference between the world price and the local market price to a special fund used to finance community activities. This solution might have to be imposed by the Ghanaian government if it is to be uniformly adopted by foreign purchasers. Yet, such a solution is not consistent with the free market policy the Ghanaian government has adopted as part of its structural adjustment programme. Such a solution would also mean that the producers would not reap the benefits of their work, although the community would benefit.

More threatening to local production is the interest shown by various Ghanaian businessmen in this trade. One company, Sabary Enterprise, is securing first contracts on buying butter from the market or employing women to process it at the lowest possible prices. They are considering setting up their own production process with their own machines and, as soon as possible, establishing a factory. If they do this, there is a real question about whether the local women's production could compete. Given the free market in Ghana and the policy of encouragement to local businesses, the future of the women's local production is somewhat uncertain. More organization and slightly more technologically sophisticated production might mean a lowering of prices for the women's product as the amount produced increases. Of course, the businessmen's production, just as the women's, depends on the availability of shea nuts, which, in turn, is dependent on the number of trees (which take fifteen years to mature) and on the climatic conditions.

Overall evaluation of the project, however, although it may fault the lack of co-ordination, must be generally positive. This project has managed to support a traditional production process through a collaboration of many Ghanaian and foreign agencies in a fashion which has substantially increased the income of village women. Marketing problems may be resolved, at least for the short-term future. Technology is continuing to be adapted in response to the problems the women are encountering. The higher technology processing by a Ghanaian business is a threat, but this is the future of many simple adaptations to local technologies which, by and large, will be bypassed as the economy develops. In the meantime, the women have learned from their processing and may be able to continue to process for local consumption and provide processing services to other women for the foreseeable future. Other options may be open to them when this enterprise becomes less promising. For the time being, this is a viable project which, at a relatively low cost, helps women increase their income and can, subject to some technical support, be maintained and run by the women themselves. In terms of measurable impacts, moreover, indications are that with more time (and assuming the technical difficulties are resolved), the women will experience positive and strong changes in income, time use and family status.

Chapter Eight: Statistically Significant Survey Analysis Results in Ghana

1) Sample type (participant, control) and literacy training (none, some)
   Chi sq. = 10.176, Phi =.405, P. = .0014
2) Sample type (participant, control) and membership in a cooperative (does not belong, belongs)
   Chi sq. = 50.632, Cramer's V= .926, P. = .0001
3) Sample type (participant, control) and membership in women's economic group (no, union, co-operative or credit association, social or religious group, combination of economic groups)
   Chi sq. = 27,012, Cramer's V = .626 , P. = .0001
4) Sample type (participant, control) and family income (high, middle, low, below subsistence)
   Chi sq. = 4.583, Cramer's V = .26, P. = .1011
5) Sample type (participant, control) and husband's ownership of livestock (none, some)
   Chi sq. = 4.176, Phi = .276, P. = .041
6) Sample type (participant, control) and respondent's ownership of land (none, some)
   Chi sq. = 4.583, Phi = .26, P. = .0323
7) Sample type (participant, control) and respondent's ownership of other assets (none, some)
   Chi sq. = 6.076, Phi = .295, P. = .0137
8) Sample type (participant, control) and change in economic activities (no change, change and less income, change and more income)
   Chi sq. = 18.261, Cramer's V = .78, P. = .0001
9) Sample type (participant, control) and education of fifth child (none, primary, more)
   Chi sq. = 7.051, Cramer's V = .469, P. = .0294
10) Sample type (participant, control) and change in time spent on children and household (no change, more time, less time)
    Chi sq. = 57.857, Cramer's V = .916, P.= .0001
11) Sample type (participant, control) and change in time spent on water and fuel (no change, more time, less time)
    Chi sq. = 55.941, Cramer's V = .675, P. = .0001
12) Sample type (participant, control) and change in time spent on leisure and rest (no change, more time, less time)
    Chi sq. = 56.122, Cramer's V = .915, P. = .0001
13) Sample type (participant, control) and change in time spent on education and self development (no change, more time, less time)
    Chi sq. = 48.546, Cramer's V = .878, P. = .0001

Decision-making

Type of decision	she decides		husband		family member		husband & wife	
	P	C	P	C	P	C	P	C
14) Use of income	24(60%)	8(28%)	8(20%)	1(4%)	0	1(44%)	8(20%)	19(66%)
Chi sq. = 17.62, Cramer's V = .505, P. = .0005								
15) Decision on her work	27(69%)	10(33%)	6(15%)	1(3%)	0	6(20%)	6(15%)	13(43%)
Chi sq. = 19.112, Cramer's V = .526, P. = .0								

P = Participant, C = Control

16) Sample type (participant, control) and change in decision-making on income (no change, respondent decided before and husband now does, husband decided before and respondent now does, other family member decided before, respondent and other family member decided before)
    Chi sq. = 17.67, Cramer's V = .514, P. = .0005

CHAPTER NINE
Tanzania – Food Processing

Climbing up the development ladder

Tanzania, with a population of 24.5 million people in a land area of 945 000 km^2, has endured, almost since independence, a severe economic situation. Economic growth overall was slow – a 2.8 per cent growth in GDP between 1980 and 1990; and it had an increasing population with an average fertility rate of 6.6 from 1965 to 1990. The overall economic growth rate from 1965 to 1990 was a negative 0.2 per cent. The Tanzanian inflation rate between 1980 and 1990 was 25.8 per cent, more than double what it was between 1965 and 1980, but average per capita annual income in 1993 was only $110. The negative balance between economic growth and fertility does not bode well for the short-term future prosperity of the country.

The Tanzanian economy is still predominantly agricultural; 59 per cent of the GDP in 1990 was from agriculture, which is an increase of 13 per cent from its contribution to the GDP in 1965. Only 12 per cent of the GDP comes from industry and 29 per cent from the service sector, which, in the latter case, is a decline of 11 per cent from its contribution in 1965. The industrial sector did not grow at all between 1980 and 1990. The service sector grew only 1.3 per cent, but agriculture in the same period grew 4.2 per cent, a relatively steady growth compared to the figures for 1965 to 1980 (World Bank, 1992:218, 220, 222, 270).

The present economic situation in Tanzania may be attributed in some measure to the political orientation of the government during the period of 1967 to the mid-1980s, however, unforeseen factors, such as the war with Uganda and the rise in the world price of oil, also had devastating effects. When Tanzania became independent in 1961, the overall policy towards economic growth was relatively conventional. The government emphasized rapidly increasing the GNP and achieving national self-sufficiency predominantly through a market economy. In 1967, however, the Government of Julius Nyerere adopted the Arusha Declaration which spelled out its new intention to move away from private enterprise and towards a government-dominated, socialist economy. Following the Arusha Declaration, the Government nationalized important sectors in industry, commerce, mining and crop marketing. Prices and wages were controlled by public organizations. Credit was rationed, and major sectors of trade were allocated to parastatal organizations. Parastatals in Tanzania proliferated; by the mid-1980s there were 425, far more than any country of its size and range of GDP (Bagachwa, 1993:91).

Initially, in the years following independence, the economy of Tanzania showed considerable promise. Between 1965 and 1976, industry and agriculture grew at healthy rates (5 and 6.5 per cent per year respectively). Furthermore, Tanzania's investment in its human resources had shown some very positive results. Literacy rates had gone from 10 per cent in 1961 to 60 per cent in 1977, by which time primary school enrolment was up to 90 per cent. Higher education had increased significantly. Life expectancy had markedly increased and infant mortality dramatically decreased. By the late seventies, however, the economy had begun to weaken substantially. Real GDP growth declined sharply, per capita income growth dropped to a negative figure, inflation soared and the budget deficit rose by more than six times between 1979 and 1985. Many factors caused this decline, among them: worsening terms of trade, an increase in the world price of oil, spending on the war with Uganda, and the break-up of the East African Community which had protected Tanzania's market. In addition, in the early seventies and twice in the early eighties, Tanzania suffered major droughts. Internal policies had also contributed to the situation. The government had invested little in the agricultural sector and had emphasized capital and import-intensive industry which led to foreign exchange shortages and other problems. Rural farmers were forced to move into villages, leaving their former lands and traditional homes. The disruption caused by this 'villagization' policy (aimed at facilitating the distribution of government extension and technical services to the farmers) caused a major loss in agricultural production, at least in the early years. The huge size of the public administration (resulting from government control of all aspects of the economy) was very costly. Observers add that the public sector was expanded 'beyond its technical and managerial capacities which has invariably been associated with proliferation of unproductive bureaucracies, financial losses, shortages and growth of second economy (parallel markets)' (Bagachwa, 1993:92–4). By 1990, Tanzania's external debt was almost three times its GNP (World Bank, 1992:264).

In this state of crisis, the World Bank and the IMF stepped in to persuade the Government of Tanzania to adopt, in 1986, a programme of Economic Recovery (the ERP). As in Ghana, this involved reducing government control of the economy both directly (through parastatals) and indirectly (through pricing and other financial mechanisms), and increasing incentives to the agricultural sector. Specifically, the government all but eliminated its categories of commodities, the trade of which was restricted to parastatals, and also its controls on prices. Agricultural marketing was liberalized, although the marketing boards and co-operative structures were not eliminated. Previously, crops could only be sold through co-operatives, now farmers were free to sell to any buyer including co-operatives and private traders, and all traders were allowed to export agricultural commodities, a privilege previously restricted to the government agencies. Private traders were now allowed to compete with the state organizations providing farm inputs such as seeds and fertilizers.

The economy responded positively to the ERP. Overall production of food and cash crops recovered significantly and growth in real per capita income overall was positive. However, Tanzania has not yet overcome all its problems. Crop-marketing problems persist, the infrastructure (roads, railroads, communications systems on which the economy depends) is in an advanced state of deterioration, inflation is still high and the growth rate slow. The lack of social services for most of the population is a major problem (Bagachwa, 1993:94–9).

Like Ghana, interest in the development of small enterprises became particularly pronounced once the era of structural adjustment had begun. In principle, the government had encouraged small enterprises almost from the beginning of independence. In 1966, the government had established the National Small Industries Corporation as a subsidiary of the National Development Corporation. In 1973, in an effort to increase its support, the Small Industries Development Organization (SIDO) was established. SIDO was supposed to promote small industries and provide assistance in marketing research, technology support and extension services. The programme for economic development, called the Basic Industrial Strategy (BIS) of 1975–90, also mentioned supporting small industries. Basic industries were to be established, including large and medium-scale industries at the national and district levels and small-scale and handicraft industries at the village level.

Little attention actually was paid to the small-scale industries in the implementation of the BIS. SIDO, however, did provide numerous services to small-scale industries. It set up industrial estates (where land, water and power were available to entrepreneurs), it made loans of equipment and capital, it facilitated technology transfer through an Indian and a Swedish technology-assistance programme and it provided extension services and training to entrepreneurs. SIDO's impact was mainly in urban areas, in large part because of the lack of resources and personnel in the rural areas. In addition, many of its programmes were conducted in the post-ERP period. Indeed, between 1985 and 1989, SIDO had prepared 609 feasibility studies, carried out 82 market surveys and trained 1427 entrepreneurs in various programmes (Bagachwa, 1993: 103). Until the ERP, the overall policy environment for small-scale enterprises was negative. The government emphasized large-scale, centralized and monopolistic parastatals which pre-empted small-scale enterprise development. Two examples illustrate the problem: in the first case, up to 1984, the state-owned National Marketing Corporation had a monopoly on obtaining and distributing raw grain and milled products. This prevented the spread of small-scale custom mills. Similarly, the establishment of a highly automated large-scale bread factory in Dar es Salaam undercut the business of thousands of small-scale bakeries (Bagachwa, 1993:104). Other policies such as the Foreign Investment Protections Act and the Nationalization Act discouraged external investment, while the emphasis on communal, as opposed to private business development – illustrated by such laws as the Ujamaa Village Act which prohibited individually owned businesses in the villages – was a major deterrent to the growth of small-scale enterprises. Nor was credit, controlled by government agencies, made available to small entrepreneurs.[1]

The ERP signalled a new era for the development of small enterprises in Tanzania. Private industrial growth and the importance of the relatively free play of the market underlay numerous reforms which facilitated private small-scale undertakings. Major among the positive reforms were foreign exchange liberalization schemes and import liberalization. Raw materials, tools and spare parts, for example, were now quite available to the entrepreneur. But, in contrast to the Ghanaian case, the support to small enterprises was not complete. Many of the old government policy biases against small-scale and private enterprise remained. Not all parastatals were dismantled and new policies continued to favour the larger employer or producer. Furthermore, import liberalization also hurt some small producers by providing more competition than their fragile businesses could tolerate. However, small enterprises did begin to grow in number and in per unit production. The Tanzanian Bureau of Statistics estimated in 1989 that the number of small microenterprises in Dar es Salaam alone was three times the level of the mid-1980s. A World Bank survey in 1990 indicated that microenterprises had on average doubled their employment levels between 1984 and 1990, while a 1990 survey of 79 microenterprises which had received SIDO support showed that most of them had experienced an increase in demand for their products (Bagachwa, 1993: 110–13).

Small-scale enterprise development did not have the same success in Tanzania as in Ghana. This appears to be in part due to current government policy as discussed above, but two other interesting factors have to be taken into account. The crisis in the formal economy (which stimulated the adoption of the ERP) did allow Tanzanian small enterprises to grow in number and expand. But they started from a much smaller base. In the first place, 'there was much less of a tradition of artisan production in Tanzania

---

[1] See additional details on licensing, registration and taxation restrictions on small enterprises (Bagachwa, 1993:104–10).

than in Ghana, so the accumulation of experience and techniques in the use of small-scale technologies and local materials had not developed to the same degree' (Dawson, 1993:74). Secondly, tight government control in Tanzania had prevented the development of a black market through which entrepreneurs would be able to get needed equipment and inputs as their Ghanaian counterparts had been able to do. In the ten years prior to the ERP, Ghanaian entrepreneurs had made substantial headway in establishing and expanding this sector, in contrast to the situation in Tanzania. The ERP helped small enterprises in both countries although, also in both cases, there were negative aspects. The results were less impressive in Tanzania and the overall picture less encouraging. After all, 'state enterprises still enjoy an effective monopoly over much economic activity, particularly in the agricultural and food processing sectors' (Dawson, 1993:74).

Poverty and prospects: Tanzanian women

Traditionally, women in Tanzania in general have not had the same degree of economic independence as their Ghanaian counterparts. Having stated this, the generalization must be quickly qualified, for there are strong variations among different ethnic groups in Tanzania as to the degree of autonomy and power a woman has in her family and village. In regard to access and ownership of land, the main asset in an agricultural economy, for example, some women such as the Luguru have the same rights as men, while others such as the Haya have no such rights at all (Fortmann, 1982:193). All rural Tanzanian women (and 87 per cent of Tanzanian women live in rural areas) have difficulties in gaining access to labour as compared to men. Where men control women's labour and can increase the amount (by adding a wife for example), women have no such outlet and can only work in collective groups for limited purposes. Women organize and hire labourers outside the family in Ghana on a more regular basis. Tanzanian women have a significantly harder time than men in getting credit for any undertaking and, as the country has turned to cash-crop production dominated by men, women get little of the proceeds despite their work in the cash-crop fields (Fortmann, 1982:193–95). Women in Ghana also suffer from male control over cash crops, but are more likely to be paid in cash, kind or return in service for their work than their Tanzanian counterparts. Women have a harder time than men in getting credit in Ghana as well.

Pinpointing the real or major difference in the traditional setting for women in Ghana and Tanzania is difficult and cannot be accomplished here; the many factors which make one country, and not the other, favour or support women are so varied. In both cases, some groups are matrilineal and some patrilineal; some groups are Muslim while others belong to other religions found in both countries. Women have differing agricultural tasks, again depending on the ethnic group in both settings. Perhaps the only factor that can be safely advanced which directly relates to this study is the traditional importance of small enterprises and trade in Ghana, in which activities women had a significant, if not a major or dominant role, as opposed to Tanzania, where there was not the same emphasis. As a result of this and other factors, women in Tanzania had traditionally less economic independence combined with fewer economic opportunities than did women overall in Ghana.

There is a certain irony in that the situation for women relative to men was and is more difficult in Tanzania than Ghana, but difficult in both cases – yet, it was in the more difficult case that the government took an explicit aim at rectifying the situation. Julius Nyerere, the founding father of Tanzania, is often quoted as saying that Tanzanian women work harder than men and do not get equal rewards for what they do, and that the Government of Tanzania intended to equalize the situation (Fortmann, 1982:191; Wiley, 1985:166). The government adopted specific laws and policies to implement this goal. Women were to receive rights equal to men, economic as well as political. Women were to receive equal education and specific laws guaranteed them equal pay for equal work. Both the national political party and the government established women's programmes in an effort to politicize women and acquaint them with their rights and potentialities. When the villagization programme was adopted, one justification of it was to improve the situation for women by 'reducing the distance to water points and offering better childcare and educational facilities' (Wiley, 1985:166). In certain family-rights areas, the government tried to support women's issues such as allowing women to keep their marriages monogamous, setting a minimum age for marriage and providing some inheritance rights for widows. The government encouraged the formation of a national women's organization, Umoja Wa Wanawake Tanzania (UWWT), structured like the national political party with cells at the regional, district, division, ward and village levels. Umoja Wa Wanawake Tanzania's mission was to mobilize women, defend their rights and stimulate their economic and social development (Wiley, 1985:167).

Despite the generally positive national government policy background, Tanzania did not eliminate discrimination against women in the workplace or in society in general. Even in the Ujamaa villages (those established in 1967 as part of the forced move of rural dwellers to villages), which had meant to favour women, long-term strategies did not include consideration of women's needs. 'Officials were reluctant to promote women's projects beyond childcare or domestic-related activities' (Wiley, 1985:168). Women in Tanzania were still primarily recorded as 'family labour' in statistical accounts, although they made up more than half of the agricultural workforce. This classification reflects female status in a 'male dominant system of property ownership and/or control' (Mbilinji, 1989:222). Women were regarded as 'not particularly bright, not capable of learning modern agriculture' (Fortmann, 1982:191–92). Despite government laws calling for equal wages, women were paid significantly less than their male counterparts. This was true in agriculture, where women were often classified as less skilled and paid accordingly. In the wage sector this differentiation was clearly demonstrated; surveys show that in the 1980s women earned less than men in the same occupations even when they had the same level of education. Moreover, as the economy of Tanzania worsened, so did the position of women. They began to be squeezed out of paid jobs and, where employed, their salaries relative to males' dropped further. In rural areas which had received no investment and began to lose the government subsidies previously provided, women had to find employment to feed their families, but they had few alternatives. Women were forced into casual labour, beer brewing, periodic migrant labour, petty trade, and other forms of periodic commodity production/trade. In this manner, women were being driven into economic activities customary to Ghanaian women and this was accepted by their male relations. The whole process was played out against a backdrop of severely increased poverty and misery. Some scholars describe the process of breaking off from the family system as the 'proletarianization' of women, in which they would then have the freedom to sell their labour and use the proceeds as they wished (Bryceson, 1985:128–52). However, the decline of agriculture and the resultant general suffering in the pre-ERP period have to be balanced against the beginnings of women receiving greater access to the economy (Mbilinji, 1989:241–4).

Following the United Nations Decade for Women conference held in Nairobi in 1985, the Tanzanian government placed more emphasis on meeting women's needs. A Ministry of Community and Women and Children was created. The introduction of the ERP in 1986 also heralded increasing concern (strongly supported by foreign donors) for the situation facing Tanzanian women. During the first period of the ERP, the overall situation in rural areas improved (although government parastatals still dominate agriculture). The situation of women was ameliorated. In the period of the second ERP, 1990–95, the government's Economic and Social Action Plan spelled out specific guidelines to increase the role and status of women in their communities. Major inequalities between men and women in Tanzania still remain, although some government-sponsored programmes have made marked advances. In education, for example, girls are just as likely as boys to receive primary schooling and, at the secondary level, girls receive less schooling but their ratio to boys has improved dramatically since the independence period. Older women are significantly more likely than men to be illiterate (80 per cent versus 46 per cent), but women aged 15 to 24 are much less likely to be illiterate (46 per cent were) and this is only 26 per cent less literate than males of the same age (UNIFEM, 1991:51). Women are gradually moving back into wage-paid jobs and increasing their percentages in various branches of commerce, industry and service. However, the overall decline or contraction of the economy is reflected in statistics on women's employment which show them substantially less likely to be in the wage-labour force in 1990 than they were in 1970 (UNIFEM, 1991:105).

The Tanzania Food Processing Project

The Women and Appropriate Food Technology Project (WAFT) was established in 1987 to help meet the needs of rural women. The project was the creation of the Tanzanian government through its Ministry of Community Development and Youth and Sports. Later, charge of the project was given to the Women's Affairs and Children's Division. The project was established as a collaborative effort shared by the Tanzanian government and UNIFEM, with assistance from UNICEF and the United Nations Volunteers. UWWT, the national women's association, was involved at the outset of WAFT to help through its work with the village women's projects. Funding was to be provided primarily by UNIFEM which would play a key role in setting up the programme, advising government authorities and maintaining accounting and technical supervision. WAFT's long-term objectives were:

1) To improve the socioeconomic status of women and to promote women's full participation in the development process, at the same time contributing to the country goal of self-sufficiency in food.
2) To improve rural women's food processing activities through the dissemination of successful low-cost technologies with a support system including training in operation, maintenance and basic management skills as well as provision of credit through a revolving loan fund.

The short-term objectives more specifically related to the ways in which the project would assist the women participating in the programme:

1) To disseminate food processing technologies that can reduce the work burden (e.g. hand-operated maize milling machines or (conditions permitting) electrically driven milling machines and other devices such as equipment for threshing, winnowing, shelling, oil processing and milk separating).
2) To train project members to operate and administer their project.
3) To prepare village women to assume decision-making positions within the village committees. Originally the project also intended to improve – increase and make more effective – the political participation of women in the villages.
4) To ensure that women have access to credit so they could obtain the agricultural tools and improved equipment for food processing.
5) To promote and reinforce the exchange of experience between women.

Ultimately, the project was expected to have multiple impacts on the lives of the women who participated. In the first place, it was intended to increase their income while decreasing the cost of processing. Credit and training would increase the chances of women starting other income-generating activities as well. Beyond income, the project was expected to improve the education of the participants, their general knowledge and skill level, their nutrition, housing, health, clothing, time budgeting and overall relations with others in the community. Basic management training through the project would provide the skills the women needed to run an income-generating project. Women would acquire, as a result, stronger self-confidence, satisfaction, and initiative in decision-making (in village committees or women's associations or political or policy groups and on community and national plans or programmes). Women would, as a result of the technologies introduced, spend less time on housework, have more leisure time and spend more time on economically productive activities.

WAFT was to be implemented through village women's groups. In the pilot phase, 110 villages in the three regions of Dar es Salaam, Coast and Lindi were to be reached. Each of the village women's groups had ten to thirty members. Group officers included a chairwoman, a secretary, a treasurer and, sometimes a miller. All group members and officers were women. Each group had its own set of regulations which specified the tasks of the members, and the principles underlying the operation and management of each group activity and the procedures for profit sharing. WAFT was to provide credit to these groups through a revolving loan fund which would charge a 14 per cent interest rate. The proceeds from each enterprise activity would be broken down into repayment of the loan, operational expenses and savings. WAFT would provide technical support through its project staff.

The three initial regions were chosen because they had few organizations which provided assistance to women. Women participants could be aged from 15 to 50; the only criteria for their selection were their participation in village women's groups and their responsibility within their household for food processing and preservation activities.

One of the key differences between this and other projects discussed is the emphasis, introduced in this project by UNIFEM, on participatory research as a method of mobilizing women, involving them directly in the formulation of project objectives and in the direction of project implementation. The other projects in this study resulted from the request of a specific group of people (cashew nut processors), or from locating such groups according to their skills (such as processing shea nuts or producing wool), or from mobilizing people because of their crying need and lack of options (silkworm raising and coconut processing). In these other cases, project staff spent considerable time trying to develop a relationship of trust and confidence with the participants. Project goals and objectives were explained and demonstrated but the project itself was developed outside the recipient community. Thus, project development and implementation was necessarily top down.

Since virtually all rural women in Tanzania are involved in food processing and most are, or have been, in some kind of women's group (group formation and group responsibility having been a major theme of the Tanzania Government at least since Arusha), need is the real criterion for this project. However, the village men and women were to be helped to study their own situation and decide what should be done to improve it. They were to select, based on their own study of the village and its resources, exactly which products should be produced and which technologies should be chosen. They would prepare or design the project and they would be responsible for its implementation. The project, therefore, would be theirs and not a 'foreign import' imposed from the outside. From the very beginning, they would be in control of their own development process. The role of the project staff would be to facilitate, and to provide technical expertise when necessary. Community Development Assistants (employees or government extension agents) would go out to the villages to help a village research team comprised of both men and women conduct a survey of the socioeconomic and technical resources and the needs of the village. Thereafter, their findings were to be presented to the village and the women's groups. The village would then establish its highest priorities and the women's groups would develop them into one or more productive projects of which they would then take full charge.

In order to carry out this scheme, a consultant from the Philippines was brought to Tanzania to hold a two-week training seminar for 35 community development workers, agricultural extension workers and representatives of the women's organizations, in the skills and knowledge needed to undertake the participatory approach with the villagers. Participants in the seminar were taken to villages for practical training and to gather data on the general socioeconomic and cultural framework of the region as a guide to project implementation. The seminar participants also learned methods for the preparation of viable projects. Project strategies, and the problems of and requirements for women's participation were also discussed. Participants, thus, received theoretical and practical preparation for the implementation of the WAFT programme. The seminar trainees were responsible for carrying out consultations with the village leaders through the Village Government Councils to make them aware of the WAFT approach and methodology.

Once consultation had taken place and research was completed, the project proposals were to be drafted with the help of the Community Development workers. These proposals were to be scrutinized at the district and regional level and submitted for approval by the WAFT project

Executive Committee. At this stage SIDO, bank specialists and experts from the Ministry of Agriculture were to provide technical support through feasibility studies. These studies were to consist of a technical description of the project, a market survey and a financial analysis.

The Ministry's Division of Women and Children was to establish a Revolving Loan Fund (RLF) which would provide a partial loan to the village women's groups. These loans were not intended to cover the entire cost of the project, as, to do this for all the envisaged projects would rapidly deplete the project budget. Instead, communities instead were to be encouraged to mobilize their own technical and financial resources and to look for additional funds. The project would facilitate this by introducing the groups to financial institutions and other potential donors. In such negotiations, the project's RLF would act as a guarantor for projects which would not otherwise be funded. Interest rates from the RLF were to be fixed at 14 per cent with various grace periods determined depending on the projects. Credit could not exceed Tsh100 000 or $222.[2]

Project implementation responsibilities were divided in three. The participant groups were supposed to provide funds, materials or labour as their contribution to the project. In particular, in the case of poultry projects and those receiving milling machines, the women's groups were supposed to build a shed before receiving any equipment. Village communities could give land, funds or labour support to the women's group. The Centre for Agricultural Mechanization and Rural Technology (CARMATEC), a local company, and the Institute of Production Innovation (IPI), an engineering section of the university which specialized in appropriate technology, produced tools and machines for the project. The project stressed the importance of using low-cost technologies manufactured in Tanzania as opposed to imported ones. The project staff were to provide guidance and training for the women in leadership, project planning and execution, project management, bookkeeping, the Revolving Loan Fund operation, and the maintenance of equipment.

UNIFEM was the major donor for the WAFT project, contributing (by the end of 1993) $250 000. Total costs for the project were $325 800. UNIFEM funds were mainly for paying project personnel costs, in-service training, and for equipment. UNIFEM played a key role in establishing the project, and in advising national authorities and remained responsible for general accounting and technical supervision while the Division of Women's and Children's Affairs of the Tanzanian Government had the responsibility of direct implementation. The United Nations Volunteers (UNV) provided volunteers who served on the staff of the project at its outset. Some staff training was paid for by UNICEF.

---

[2] In this report, the Tanzanian shilling is valued at 450 to the US$1.

The dream and the reality

Although the original formulation for the project called for 110 villages to be included in a pilot stage, in fact only 26 village women's groups have been provided with funds (by Spring 1993) from the revolving fund. In 1988, the first gardening and poultry projects were implemented in the Coast Region. The milling machine projects were postponed at this time because of unforeseen delays in receiving materials and building the required sheds. From 1989 to 1991, feasibility studies for all villages were prepared and submitted to UNIFEM for funding. However, because of a lack of staff and equipment (most of) the projects could not be carried out. In 1992, Lindi Region received the first materials and technology for projects in the zone. No project has yet been implemented in the Dar es Salaam rural region.

One of the chief elements of the project which had to be re-evaluated and revised was the timeframe. Originally, the project was supposed to be completed in two years, after which the outside funding agencies should have been able to withdraw. This, however, proved to be far too little time and the project's first phase had to be extended to 1991. The very methodology of the project required participants to be trained and staff to be even more extensively prepared. As there was no entrepreneurial tradition in the villages, and most of the women were illiterate, time was required to help them understand what such projects could do and what possible choices there might be. By 1993 many women's groups were only just embarking on their projects and receiving their loans and equipment. The government found that it may have to establish an NGO in charge of the project activities. The staff of this Women's Development Fund would have more flexibility than those currently involved in the project inside the Division. Currently WAFT is using the remaining resources from its first funding and the repaid loans to continue to help women's groups start a project under the system devised for this programme. But the original target number of 110 villages for the first phase is clearly out of range. Projects have been established in twenty-six villages, if another twelve were added with this funding and in this phase, it would be doing well. At the time of Valerie Autissier's visit, in August 1993, 983 women participants had been helped through WAFT, and WAFT was seeking to embark on a second phase of this project.

Women's voices

In this study, a questionnaire was administered to forty participant women; thirty other women in two regions formed the control group (five villages from the Coast Region and two from Lindi Region). All the women were rural villagers. Most of both groups were aged thirty-one to fifty. Slightly more of the participants were in the older age brackets but the difference was not significant even when age categories were collapsed. A quarter of both groups said they were the heads of their household while the other women were not. The education patterns of the two groups was similar, although the younger women in the control group were slightly more likely to have had schooling. More than half the participants had not been to school while only forty per cent of the other women were in this situation. Almost all of the women in both groups who had not received education had had literacy training, a rare fact when compared with the other countries.

More than ninety per cent of both groups of women were Muslim. Most women were married – about sixty per cent – but there were a small number of divorced, widowed and even single women in both samples. Among the Muslim women, most did not have a co-wife, but participant women were significantly more likely to have one than were other women although the participants were also more likely to be the first wife when they had a co-wife. Consistently, the slightly older participant women had more children than the other women. Most women said their husbands were farmers and that they had had some or all of primary school. As far as their own principal occupation was concerned, more than eighty-three per cent of both groups said they were farmers. Most women said they raised rainfed crops; a large number of both groups also raised vegetables as well. Most of the women used their produce to feed their families although a smaller number of women in both samples said they sold what they grew. More than forty per cent of both groups said the income they earned from their various activities (including small enterprise work) amounted to fifty per cent or more of the family income while the rest said they earned a small or insignificant income – but an income nonetheless.

Almost all the participant women surveyed had joined the project more than six years before and all of them said they joined because it would help them, or their women's group, get credit and training or simply a larger income. In fact, in their eyes the project did not appear to have had a major impact on their income. Ninety-three per cent of them said it had not increased their income. The same number said the project had not affected their economic activities significantly. Their disillusionment in this regard is reflected in the lack of difference in family income between them and the other women. Most women fell into the low-income category while about a third of both groups lived below the poverty level. Almost a quarter of the non-project women, however, were classified as middle income compared to only thirteen per cent of the participants.[3] This profile is perhaps less

---

[3] The scale used was: annual income below poverty = Tsh249 999 or less; low income = 205 000 to 209 999; middle = 210 000 to 479 999; high = 480 000 or above.

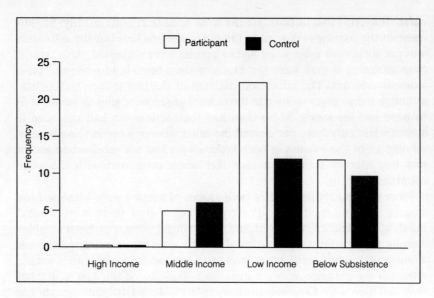

Figure 9.1 *Income levels*

disturbing from the point of view of project impact since the non-project women were also significantly more likely to receive income from relatives who had migrated out of the zone to find work. Nonetheless, the results did not indicate a strong contribution from the project, supplementing the income of the beneficiaries.

There were some significant differences between the participants and others in regard to family assets. While almost all of both groups reported their husbands owned land and a house, more of the non-project women said their husbands had livestock while more of the participant women said their husbands had some other asset. More than three-quarters of both groups, however, said their husbands received no income from these assets. Nor were there significant differences between the groups in regard to the woman's possession of assets. Most owned a house or land, most did not own livestock, and most did not own other assets. Ninety-five per cent of both groups said they received no income from whatever assets they had. Nor did either group claim to have acquired their assets during the period of the project. In fact, non-project women were more likely to have done so. One hundred per cent of the participant women said the project had had no effect on their acquisition of assets.

Most in both groups spent their money either on maintaining the house or on food, with the participants more likely to buy food. None of them felt this expenditure pattern had changed much over the last few years. Only one women said the project had had any impact on this at all. There was no real difference between the two samples in regard to family diet; more than eighty per cent had regular access to meat and other good staple foods. Only four participant women said the project had helped them get better food for their families. Nor was the education level of the participants' children higher than that of the other women, and none of the participants attributed any positive impacts on their children's education to their project experience. There were, however, other differences in family patterns between the two groups which were significant. For example, significantly more of the participants said they now spent less time caring for their children or their household. They were likely to say they had less time for leisure activities or rest than they used to have, although this difference was not as marked. More of the participants also claimed to have less time now for their own education or self-development than the other women but this tendency was not great as most women had made no change in this regard. A larger percentage of the non-project women wanted some change in their time use, preferring to have more time for work (participants 13 per cent, control 24 per cent), but most women in both groups did not want to change their time use. Nor were the participants sure of the impact the project had had on their time use despite the different pattern they exhibited from the other women. Fifty per cent of them said it had had no impact, that is, they could not themselves see any change in the way they spent time now. Thirty-two per cent said it had resulted in their working longer hours and spending less time on the household and their children. The rest were unsure.

There were also differences between participants and other women in regard to family decision-making authority but, where these differences emerged clearly, the participant women had less authority. In regard to deciding the use of their income, more non-project women made this decision by themselves, although most of the participants at least participated in this decision with their husbands. The majority of both groups made their own decision about what they worked on and most participated at least in the decision on where they lived. Significantly more of the non-project women, however, made the decision on what the family would eat although participant women at least were consulted, or consulted with, their husbands in this matter. Most women at least participated in the decision on their boy children's education although approximately a third of both groups said their husbands decided this. Almost all the women decided on their daughters' education themselves. Virtually all the women said their decision-making patterns were the same as they always had been and only one participant woman said the project had had any effect on the family authority pattern – she did say it had given her more power. Here, as in Ghana, the question arises as to whether the greater economic possibilities opened to the women caused them to lose some of the independence they may have had because their husbands now take a greater interest in what they do as it promises to be a more important share of the family income. The women themselves do not perceive this pattern but it seems to be a possible interpretation which deserves further research.

More than three-quarters of the participants said the project had not affected their life. Four women said it had affected their lives by introducing them to a new technology and two said it had affected them by giving them training. Only one woman said it had raised her income. Asked directly if the project trained them in any way, more than fifty per cent said it had not, while forty-eight per cent acknowledged that it had. Ninety-two per cent of the women said the project had given them access to credit or credit and a new technology. But, again, ninety-three per cent of them felt it had not increased their income. Eighty-eight per cent of the women said the project had no effect on their family at all. One woman did say it gave her more independence but another said that it had in fact cost her money; her income had declined as a result. The same optimistic woman responded to the direct question about decision-making power in the family by saying she now had more power. The others did not feel it had any influence on their status or authority at all. There were, however, two dimensions on which the participant women were positive. They said their husbands liked the project and they also said (at least sixty-three per cent did) that the project made them more self-confident or, in some way, more optimistic about their future. They did not specify, as did women in some of the other projects, that they now had confidence in their economic prospects for the future.

Looking more closely at the questionnaires, it is clear that certain village projects had major problems and the result was a sense of discouragement among women who had participated. Thus in Biwilingu, in the Bagamayo District (Coast Region), the women's poultry project encountered numerous difficulties including diseases which decimated their stock and disagreements within the group. Participants responded to questions about the project's impact with disillusionment. One said, 'I feel worse about myself because the project failed to take off and there is no group cohesion'. Another reported, 'I feel bad about myself. I'm repaying a loss which I'm sure is a result of ignorance. I was not trained properly to keep books of accounts.' Another said the project had had no positive results, and that she considered her involvement a waste of her time, while she accused the group leaders of keeping all the information to themselves. One woman said in anger, 'Clever women in this group have hidden all the information on the performance of our group. We don't know how much we have and I personally can't tell. I don't know whether I feel better or not.'

Musugusugu in the Kibhaba District (also Coast Zone) was a gardening project which suffered from a dearth of water. Furthermore, the women in the group were generally older than the average sampled; they ranged in age from forty to seventy-three years old. The lack of water meant that the group had to carry water long distances. The women were discouraged and disparaging. One (aged 65) said, 'We have to carry water on our heads . . . I'm very old now.' Another (73) said, 'I work a lot. Carrying buckets on my head as old as I am is not good.' Another younger woman (40) who was more positive about the group nonetheless commented about the older age and lack of strength of the group pointing out that 'Youths are not ready to join the group because of the unavailability of water, and girls are not ready to carry water on their heads.' I.H. Mafwenga, who administered the questionnaires, confirms that most of the group members are old and can not cope with the water problem.

Yet, negative comments from the projects which have had financial or organizational difficulties are certainly outweighed by the positive comments from most of those surveyed, as the overall statistics reported above show. It is striking that in the absence of any substantial increase in income, most participants still felt the project had done something for them. Many made comments like the following woman from Mkuranga in Kisarawe District (Coast Region). She said, 'I feel happy. Few women own and operate milling machines, but I do!' Another woman in her group said the same, 'I feel better about myself because I also own the machine (being a member in the enterprise). Owning a milling machine is prestigious.' Another said, 'I feel happy and I'm confident that I'll benefit from the enterprise.' Another: 'I feel happy . . . I can't explain or express myself.' Others in the group made similar statements despite the fact that due to a lack of grain there had not been enough business for the group to be able to begin to repay its loan.

In Kiwangu in Bagamayo District (Coast Region), where the milling project had much greater success and the group had even been able to repay all its loan, the responses were very favourable. 'I feel proud,' said one, 'I feel proud because other women in the village look on us as fortunate fellows.' Another said, 'I feel myself superior, especially when I operate and service the machine before other women.' A third said, 'I feel proud because others (women) envy us.' In Vikindu in Kisarawe District, which had a successful gardening project, women gave additional reasons for feeling better about themselves. One said, 'I'm proud because we are the few women who have been able to get credit in our village.' Another said she felt better about herself 'because I anticipate some income in the future from the enterprise. Other women not in the group will have no hope anywhere.' A third said she got opportunities she would never have had to meet people from outside, including Europeans and government officials with whom she could have discussions. Other women, she said, would not have such an opportunity. In Lionja and Naipanga, in Nachingwea District (Lindi Region) which both had successful milling projects, women were equally enthusiastic. One pointed out that not only she but also her children had been relieved from the arduous task of pounding grain by mortar. Another pointed out that as a result of the project she now has the confidence to talk freely in front of other people. She has the confidence also to go into town with a lot of money to purchase supplies. Most women said they were 'happy' or 'proud' – owning and operating a machine set them apart from other women, gave them a skill which women ordinarily did not possess, and made them optimistic about the future. Even in Musugusugu and Biwilingu there were women who were positive. One said that training had given her more confidence among other villagers, while another said simply that she had hopes for the future because of the project.

The picture emerging is one which supports the earlier surveys in which the women are said to be generally enthusiastic and appreciative of the project. Here, as in many of our other cases, they do not see the project as having had much of an effect on their overall pattern of daily living. Nor do they objectively (through comparisons with the women who had not been in such a project) demonstrate much of an effect. Where differences exist between the women who were not involved and those who were, the measures of economic status and personal authority tend to favour the non-project group. In just one dimension, there is a strong positive characteristic of the participants not shared by the uninvolved women; participants are significantly more likely to belong to a co-operative or to another women's economic group.

This difference, indicating greater socialization in women's activities and group potential for activities, may in part be attributed to the influence of WAFT. Women were required to join a co-operative or already be in one

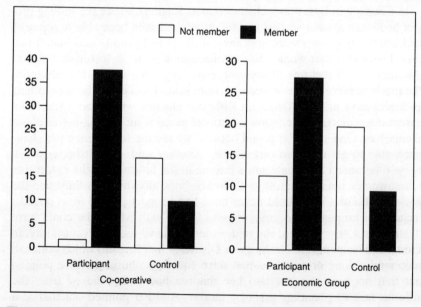

Figure 9.2 *Membership of co-operative or economic groups*

to be able to participate in the project. In the early stages, the national women's group, UWWT, did mobilize village women generally into groups as part of the preparation for this venture. Thus, some women's groups were already in existence as this had been part of the government's philosophy (grouping and co-operatives) since the Ujamaa Declaration and were not a result of any project work. In any case, WAFT reinforced and extended this emphasis on group socialization. One probable result of this socialization, combined with the training and group work involved in the project enterprises, is the sense of pride so widely shared by project participants. Women not in WAFT enterprises score better on other developmental measures such as personal independence and even income. Perhaps this is because overall they are younger. In any case, it would be a mistake to make too harsh a conclusion about WAFT's impacts from these results. Discussions above indicate numerous problems which WAFT has encountered. Because of these many problems most of the women's village enterprises now underway have only been started relatively recently, in the last year or two. The strong economic results shown in other projects emerge. Given the importance of trying to involve women in the process of decisions on the nature and form and implementation of their enterprise, and given their own pride and enthusiasm in their work, it is disappointing to find so few indications, or evidence of, the new participatory approach being effective in economic terms. A longer-term perspective should, however, provide a much more positive picture.

Realizing the dream

The Tanzanian Food Processing Project has had a significant level of success in line with its initial objectives – if the time frame is considered. Twenty-six women's groups have gone through the entire process, from participating in the research, to selecting a project and helping draw up a research design (which is tested for feasibility by outside specialists), through training in the techniques of the technologies required and in the management and administration of such a project. These groups have embarked on a project. Nineteen of them have already begun to pay back what they borrowed. Both the government and women's groups have been and continue to be enthusiastic about the project and its potential, which argues for the project's continuation and expansion.

Not all aspects of the project have worked out as intended at the outset. One problem lies in the scope and ambition of the project's initial goals. It may not be possible to do everything at once, and the accomplishment of one goal may, at least in the short run, be detrimental to another. One specific example is the dual goals of increasing income and decreasing workload, while increasing leisure time. Some of the projects chosen increase women's workloads although they also increased income (garden and tractor projects, for example). It will only be in the long run, if profits are high enough, that women will be able to gain leisure time by paying for services or buying equipment which will diminish the time required for other tasks. In the meantime, participant women seem more concerned about income than their own time use (Mbughuni *et al*, 1989:24–5).

An additional example of conflicting goals is the project's emphasis both on supporting simple low-cost technologies and on increasing income to the maximum. Participant women often chose expensive and higher-level technologies such as electrical milling machines. These certainly decrease women's working time and increase their income more than hand machines or other simple household food processing technologies, but they require a much larger subvention from the outside and may not, in many cases, be economically viable. The participant women were more likely to choose the milling machines than any other technology. Their choice is in part because of information – this was the best-known technological improvement in the region – but certainly also because, if obtained, it would dramatically increase production (Mbughuni *et al*; 1989:24–5).

The designation of the target group of the project was not precisely defined or accomplished in so far as it had been identified. Originally the project had intended to reach the poorest village women – 'the poorest of the poor' – but the selection of women was left up to each region. In addition, the project required women and their village to provide financial and material resources. As a result, although the participants were all village women, they were often from villages which did have resources already. One of the successful projects was in Kiwangwa village which already had a thriving pineapple trade which provided a cash income to many families in the village.

A third problem encountered by the project is a result of the methodology adopted by the project and the preparation of the implementing staff. Participatory research calls for staff who understand how to do such work, has internalized the participatory approach and has sufficient technical training to understand feasibility and market studies. Yet the UNIFEM-sponsored 1989 report on the project points out that the staff, especially the community development assistants, had a 'traditional framework of donor aid' in which their expectations were they would provide some activity or income to increase the family welfare. They did not evaluate proposed projects in terms of their sustainability. Nor were they prepared to do feasibility studies which could ascertain the long-run viability of a particular activity. As a result, their studies of projects proposed by the women did not include proper analysis of markets, supply of raw materials, or even repayment schedules.[4] Thus, many of the projects were either unfeasible or

---

[4] Ibid., p. 25.

not as well-planned as they should have been. In one case, for example, a woman's group chose an electric milling machine although there was no electricity in the village (Mbughuni *et al*, 25). In other cases, the capacity of the milling machinery was underestimated, resulting in underuse and over spending. The tractor projects may not be viable because of the costs of the initial investment and later running costs (if spare parts could be found at all).

Women participants did not seem to understand thoroughly the nature of the programme and their role in it. In some cases, they were not even aware of having participated in a research process to assess village needs. When they were aware of having been involved in identifying the problems facing the village, there was no necessary connection between this and the choice of technology for the project. That choice was often suggested by the community development assistants. Conversely, the women sometimes approached the project with a technology firmly in mind about which they knew little (typically milling machines), and even went ahead with building the needed shed before any research, problem designation or feasibility study had been carried out. Furthermore, the women were not trained to participate in the actual project design nor in feasibility studies and therefore had little direct information on the commercial viability or constraints on their projects.

Other constraints included lack of sufficient resources. The project budget was, of course, prepared before the women's projects were chosen. As these ran to higher-level and more expensive technologies than had originally been envisaged, the funds were quickly exhausted. In addition, the Government of Tanzania had agreed to provide additional resources at the district level, but did not. As the project did not include funds for running costs at the district and regional levels, this lack created a barrier to project efficiency. Funding was also a problem because it was delayed owing to processing demands (and sometimes just 'red tape') which meant that many village women's groups had to wait long periods before starting their project. In the Lindi Region, for example, funds were received two years after the proposals were written. This led to the discouragement of many women – in the case of some villages in Dar es Salaam, some groups have dropped out altogether and have no further interest in the project.

Another drawback was in training. Although the programmes at both the staff and the village level were held to be good and useful, staff members report that there was insufficient training; more programmes of greater length were needed. Nor was the training equally effective in all regions. In Dar es Salaam it was apparently well-planned and executed. In the other two regions training was shorter and was not scheduled close enough to the time of project implementation to be of optimum use.

Staffing was a further problem for the project. There were only three full-time national staff supported by the community development workers.

The national staff felt that the definition of their tasks was simultaneously too ambitious and too vague. The same problem was identified by the district and village staff. The result of the unclear and global terms of reference was poor reporting and monitoring, and a lack of communication between project staff and government collaborators. Staff were reshuffled or replaced several times, which led to a lack of continuity. Nor were there adequate specialized or technical staff to administer some of the project needs, especially in the areas of credit and appropriate technologies.[5] District staff were expected to provide management, training and technical advice, as well as supervision of activities, but generally did not have the training for all of this. Expert advice could not be requested from other government personnel because of difficulties in co-ordinating with them, in part because there were no allowances for travel or means of transportation. Staff also suffered from a general lack of means of transportation. For the entire project there were only four motorized vehicles and twelve bicycles to cover the huge geographic area represented by the three regions. This made staff efficiency very difficult and undermined any close supervision of what was occurring.

The provision of credit was an essential component of the project but was undercut by problems. Originally, a credit officer was to be part of the project but no one was ever hired. Credit is not traditionally offered in the Tanzanian villages so the concept of credit was very difficult for the women to understand. They were better able to understand subsidies or gifts than money which they had to pay back. They were able to comprehend that the money must be repaid to be made available to other groups, but their understanding of the process was slight. In Valerie Autissier's two village visits, the women did not know their income, much less what they still owed on their loan.[6]

Additionally, because the women in many cases chose higher-cost technologies such as electrical mills, they were dependent on large loans and continuing outside support. Although they may be able to repay the initial loan for the equipment, any unforeseen charges (and without proper feasibility studies these may be many) will require further loans which they may not be able to repay. Thus, the credit structure for many of the women's groups is rather fragile.

Despite all the problems and drawbacks encountered by the project, project staff remain enthusiastic about it. They point out that women who have obtained loans are now able to open and run bank accounts and have experience with credit institutions. The women have gained experience in working in groups and have acquired skills in areas previously reserved for

---

[5] SIDO was to provide technical support, especially in credit and technology, but did not.
[6] We realize it might be difficult for the women, to disclose their income to a 'stranger' without any reservation or reluctance.

men. Access to communication and training seems to have built up their self-awareness and self-confidence. Women report that their status has improved, they have more respect now from the village and more power in relation to men than previously. Staff evaluations show that women have become more involved in village activities, and their experience has had an impact on other women in non-participant villages who have formed groups hoping to obtain some of the same benefits themselves (UNIFEM 1993b:4–6).

Of course, the levels of project success remain very uneven. Some women's groups have taken full charge of their activity and manage their business well. In others, the enterprise has failed. The milling and gardening projects have had the highest rate of success. Milling projects have suffered from delays in project design and problems with machine installation and breakdowns. The machine itself is a serious investment for a women's group since it is worth about ten thousand dollars – village women normally handle capital of about fifty dollars. The machine is so expensive that the loan can be serviced but profits per individual are low.

In some villages with milling projects, the quantity of grain available for milling is too small for a milling project to be economically viable. In these cases it is very difficult for the women's group to repay the loan or generate any income at all. However, the provision of the mills means a grinding service is available in the villages which women see as a great advantage. The women participating in these projects are proud of them and have used their experience to embark on other ventures. Several poultry projects failed because of disease and lack of technical experience but the surviving projects are doing well. One pottery project failed because the older women with the expertise could not participate in the group activities, while the younger women dropped out to get married and participate in other family activities. The oil processing schemes failed due to inadequate technology. The gardening projects are reported to be successful although we have no direct information on them. The dairy cattle projects have not been fully implemented because of delays in equipment delivery and in the completion of the required sheds.

A summary final evaluation of the project to date indicates a certain financial fragility. Income generation is low on average although some groups are able to pay back loans. Furthermore, the relative success of the groups has stimulated imitation which is good, but private individuals with more means and expertise now threaten the future viability of some of the women's groups' projects and raise questions about the long-run future of many of the enterprises. There is some question about the efficacy to date of the chosen project methodology – participatory research. Many of the women did not understand the process at all, while many of the community development workers, not highly technically trained themselves, went through the process maintaining their overall top-down, welfare approach. As a result, projects are sometimes ill-planned and women do not always completely understand and therefore control their choices and their development progress. The Government of Tanzania has apparently adopted the participatory approach for its other projects with women. This raises some questions as the amount of staff and time needed for such an approach are greater than a traditional project and therefore more costly, while the results to date do not completely warrant this choice, at least not in terms of the initial goals set for WAFT.

On the other hand, six years is a short time for this radically different approach to be internalized either by uneducated village women or extension staff. The clearly positive side of the participatory research strategy is the emphasis on including the participant as much as possible in all levels of decision-making, from identifying the problem, through designing the solution, to taking charge of the implementation. Despite the problems in execution, this overall approach shapes the way government workers look at work in the village with village women and creates a response from the women which ultimately will include increased understanding and responsibility on their part. Finally, evaluations and reports submitted to WAFT and UNIFEM have shown a high degree of enthusiasm and interest in the project in the villages even though the women are not getting the incomes they had hoped for when they joined at the outset. This enthusiasm must be attributed to the feeling of possibility and capability or, as desired by the project, some level of self-empowerment enjoyed by the women. In a situation where Tanzanian village women have had little chance to change or improve their lives, WAFT has opened a new set of alternatives with great promise.

Chapter Nine: Statistically Significant Survey Analysis Results in Tanzania[7]

1) Sample type (participant, control) and membership in a co-operative (does not belong, belongs)
   Chi sq. = 27.778, Phi = .63, P. = .0001
2) Sample type (participant, control) and membership in a women's economic group (no, yes)
   Chi sq. = 9.287, Phi = .364, P. = .0023
4) Sample type (participant, control) and husband's ownership of livestock (none, some)
   Chi sq. = 2.865, Phi = .258, P. = .0905
5) Sample type (participant, control) and husband's ownership of other assets (none, some)
   Chi sq. = 4.57, Phi = .326, P. = .0325
6) Sample type (participant, control) and change in time spent on children and household (no change, more time, less time)
   Chi sq. = 5.645, Cramer's V = .286, P. = .0595
7) Sample type (participant, control) and change in time spent on leisure and rest (no change, more time, less time)
   Chi sq. = 5.391, Cramer's V = .28, P. = .0675

Decision-making:

Type of decision	she decides		husband		family member		husband & wife	
	P	C	P	C	P	C	P	C
8) Use of income Chi sq. = 6.33, Cramer's V = .301, P. = .0965	16(40%)	18(60%)	12(30%)	10(33%)	2(5%)	0	10(25%)	2(7%)
9) Decision on family diet Chi sq. = 9.21, Cramer's V = .368, P. = .0266	22(58%)	21(70%)	2(5%)	6(20%)	5(13%)	0	9(24%)	3(10%)

P = participant, C = Control Sample

10) Sample type (participant, control) and contribution to family income of non-resident relative (no, large, small or medium, insubstantial)
    Chi sq. = 7.626, Cramer's V = .332, P. = .0544
11) Sample type (participant, control) and existence of co-wife (none, yes)
    Chi sq. = 3.03, Phi = .223, P = .0817.

---

[7] Significantly different results are reported in the table and the full results including percentages of the significantly different decision-making comparisons are also reported. Note that usual levels of significance reported in this study are .05 or less, while several (# 4,6,8,11) of these relationships are less significant than this. They are included because the trend observed there was reported and weighed in the text.

CHAPTER TEN
A Comparison of the Impacts of Different Strategies . . .

Basis of comparison

This study has examined eight projects in terms of their effects on the income and assets, the quality of life, time use and the outlook and expectations of the men and women who participated in them. We found varying kinds and levels of impactsand this chapter, by looking across all eight cases, will draw out how different choices in strategies within the projects may or may not have resulted in the varying outcomes. However, it is not just in those cases where the developmental impacts considered in this study were the greatest that we find projects which are considered successes. *All* of the projects were chosen for study here because they had accomplished goals which were considered interesting and important by planners and evaluators concerned with facilitating the development process for women. Nor should any of these accomplishments be undervalued. It is very difficult to bring about change in impoverished and traditional environments.

Furthermore, overall criteria for comparison other than those chosen here are extremely important. One of these is sustainability or replicability; that is, could these projects last at least over the income-earning life of the women involved and would they spread to other women who had not initially been included in the work? This combined standard is extremely important for those who are calculating the best use of the limited monies available to help in poor countries. Unless the projects have a long term result *and* unless they go beyond benefiting a few individuals, they may not be worth the investment of the scarce resources (in human as well as financial terms) available. An alternative, but related, standard could be the achievement of a long-term commitment by the national government to the methods and strategies adopted in the project. This accomplishment would guarantee the longer-run and more widespread impact of the project's goals as the government would incorporate these into its own plans and programmes. Balanced against these desirable outcomes is initial cost. Where the projects are exceedingly expensive, they may be beyond the means of a specific donor or group of donors or national agency no matter how good are the outcomes they promise.

Table 10.1: Comparing project results

	Peru	Honduras	Guatemala	Bangladesh	India	Thailand	Ghana	Tanzania
Sustainability (evidence that will endure)	low	low	high	low	high	low	high	low/medium
Replicability (evidence that will spread)	low	low	high	low	high	low	medium	medium
Government commitment to project	low	low	medium	low	high	low	medium	high
Started use of new technology	yes	yes	yes	yes	yes	no	yes	introduce technology new to region
Began new enterprise	no	yes	no	yes	yes	yes	no	yes
Began new local NGO	no	yes	no	no	no	no	no	no
Began new producers/marketing co-operative	no	no	yes	no	no	no	no	no
Opened new market	no	yes	yes	yes	yes	no	yes	no
Number of direct beneficiaries at project end	250	158	465	108	500*	406*	449	983
Years project lasted	3	7	5	13	10	2	4	4
Years since completed	2	5	1	n/a	1	3	n/a	n/a
Project cost** (US$000s)	376	200	903	100	530	40	150	41

* In India, 2500 other women and their families directly benefited although they were not technically in the project. In Thailand, the figure represents those who worked in the factory or benefited from the investment funds. The farmers who sold to the factory also benefited and they number approximately 700.

** Because many local agencies' records were not complete or not available, our figures on project costs are estimated based primarily on what the international agencies that sponsored them had recorded. The figures should only be used to compare the relative size of costs as they are far from exact. In Bangladesh for example, the value of considerable volunteer time is not entered, nor are the overhead costs of MCC.

In Table 10.1 above, we compare the performance of the eight projects according to these standards and show also the other accomplishments which their managers and/or sponsors felt were criteria for judging them to be successful.

Looking at this table, then, we find that, except in the case of Thailand, cost seems to relate to numbers of people reached and possible replicability, as does extent of government involvement. The Thai project, however, illustrates the issue which forms the centre of this study. At the height of the project, during a two-year period, as many as 1300 people may be said to be benefiting (from higher salaries than otherwise available, or from a new and better-paying outlet for their crops) from the venture capital project. This benefit did not, however, last long and appeared to have no discernible impact on their lives or work beyond the specific brief period in which the ginger-pickling factory was open. We are drawn back, then, to asking under what conditions *long-term* developmental impacts ensue.

At the outset, we emphasized the importance of context both in terms of national and regional (local) economic development together with government support for the informal sector, and the access women in the country (and the specific project region) have to the economy generally. We hypothesized that impacts would be more difficult to achieve (require more intervention with more complex packages of strategies) in the most unfavourable contexts where national growth was slow and government support weak, and where women had had the least access to the economy. Where a successful intervention took place in such a disadvantaged region and a sustainable enterprise was established, the impacts might be greater because the participants were starting from a much less privileged base. These assumptions must now be examined in the light of the results of the data so far presented.

What emerged in the eight cases was the following set of developmental impacts:

Table 10.2 shows that most of these projects did have positive developmental impacts for women beneficiaries. The two which did not were those in which a one-dimensional intervention programme was emphasized. Thus, in Peru the emphasis was on developing and introducing new technologies, while in Thailand the project only made investment capital available. This finding suggests support for the hypothesis advanced above: that multi-faceted projects had greater and further-reaching impacts. However, what this means is unclear until we take into account the level of development of the country, women's access to the economy generally and variations in project participants' abilities to receive projects such as age, education, and marital status. The section below discusses findings from a quantitative analysis of all impacts across all the projects taking the contextual variables and individual characteristics of project participants into account.

Table 10.2 Comparing project impacts (improvements achieved according to survey responses)

Country	Income	Assets	Mobilization (group takes responsibility in community)	More Time for rest	Family Quality of Life	Authority in Decision-making	Status	Self-Confidence
Peru	no	no	no	no	no	no	no	no
Honduras	yes	yes	no	no	yes	yes	yes	yes
Guatemala (wives)	no	no	no	yes	yes	no	yes	no
Bangladesh	yes	yes	no	no	yes	yes	yes	yes
India	yes	yes	no (yes)	no	yes	yes	yes	yes
Thailand (factory workers & farmers)	no	no	no	no	no	no	no	no
Ghana	no*	no*	yes	yes	no*	no	yes	yes
Tanzania	no*	no*	yes	no	no*	no	yes	yes

*The projects in Ghana and Tanzania were too recent to judge the long-term impact on income, assets and therefore quality of family life.

Table 10.3: Variables

I: Independent strategy components, project and participant characteristics		II: Dependent developmental outcomes: ('performance' or 'impact')
Independent	Control	
Mobilization	Level of development	**Decision-making power (her, husband, both)**
Training	Age of participants	Over woman's income
Access to credit	Education	Over if she works and at what
Technology introduction		Over family diet
Project age		Over boys'/girls' education
Marketing		
		Time-use changes
		Spent with children and on household
		Spent on own education and self-development
		Economic status and family well-being
		Perception of overall change in economic activities
		Economic activities (change in)
		Perception of overall change in economic activities
		Social/economic activity
		Membership of a co-operative
		Perception of project impact
		Improved her life in major ways
		Increased her income
		Major positive impact on her family
		Gave her more authority in her family
		More self-confidence and made her feel better about her economic future

1 Many dependent variables could also be viewed as independent, that is, existent cultural norms define what rights, responsibilities, and choices a woman has, including how much authority she exerts in family decisions, what assets she owns, and even how she spends her time. However, following the developmental goals outlined above, we are exploring in our study of 'impacts' *changes* which increase authority or ameliorate conditions in women's daily lives, and these are seen as dependent variables.

Comparative data analysis

In this section, we look at the actual relationships among the strategy components (our independent variables), together with the control variables (measures of context and individual characteristics), and the measures of developmental impact on the women participants in the projects (the dependent variables). Guatemala is eliminated from the discussion because it targeted male participants and women were only added at a late stage and not even to the central part of the project.

We have focused on eight independent and control variables and seventeen dependent variables classed in five groups. Provision of initial capital, although interesting in theory, is provided in two cases, once in place of credit, another time along with it. After preliminary tests, it is excluded because of its lack of general explanatory power. We have also omitted our construct variable 'access of women to the economy' because initial tests suggest it does not explain results, perhaps because the construct is not closely enough related to the actual status of women. We also do not consider the choice of skilled or unskilled women in the comparison since only in Ghana were skilled women chosen (given that Guatemala is eliminated from this analysis).

For our data analysis, a variety of statistical tests were used. Contingency tables were constructed and Chi Square and Kendall's tau were computed to test for independence where appropriate. An analysis of variance (Anova) was carried out to test for significant group difference between certain attributes (i.e. control versus experimental, membership in a co-operative or no membership etc.). A factor analysis was also performed on all 'performance' or 'impact' variables using a varimax rotation. Finally, regression and correlation analyses were carried out. The following subsection discusses the key findings derived from these various statistical computations.

Findings

First, although the individual country cases clearly indicated differences between the participants and the control women, we have taken the entire group of both samples from the seven countries and tested the extent to which significantly different patterns are found in selected changes in lifestyle through Anova tests. If the projects overall had impacts which go beyond 'normal' patterns of change in each country, than distinct differences between the two samples should emerge. In fact this proved to be so, although some differences may have been due to project selection criteria rather than being an impact of the programme introduced. We found, for example, that participant women were significantly more likely to belong to a co-operative and thus be involved in group activities and group socialization. This of course, as a pre-condition for many of the projects studied, does not reflect project experience, but does indicate a difference between the experimental and control groups. Women who had been in projects, however, were also significantly more likely to have changed their use of time over the project period so that they had less available for their children and the care of their household (and more for their work) and also had less time to rest, both of which results may be inferred to be project impacts. Participants were also significantly more likely to have changed their use of time in order to include more time for their own education and other development activities. They also had significantly higher family incomes and were more likely to have assets (beyond land or a home or livestock).

In regard to decision-making, most women decided on how to spend the money they earned, although many consulted their husbands and did not decide alone. There was a difference, however, in deciding whether they worked outside the home and what they did in their work. Participants were significantly more likely to make these decisions, while other women had a much greater tendency to be under the control of their husbands. (Patterns in other decisions seemed to relate more to cultural factors rather than to project participation.)

The overall sample, then, does show participants with some important changes in their lives not shared by women who had not had access to one of the projects. Most of these impacts are positive, empowering women and improving their lives. However, this is not so in regard to lessening the load women bear. We find participants are working more and not less. This is a cost to their families in terms of the time they have available for them and it is a cost in terms of their own labour – they have less time to rest.

Next we explored the entire group of women participants in regard to the one-on-one relationship between the factors which might cause differences in impact (the independent variables) and those indicating developmental level, change therein, or perception of project impacts on these things (the 'impact' or dependent variables). Here, we find some interesting relationships, a few of the more important of which are reported here. The most numerous significant relationships emerge with three components: age of the project, training (intensive training in the use of a technology and sometimes in the management of credit or other business skills), and emphasis on marketing.

Women who are in older projects are more likely to say that the project provided them with a higher income and changed their life in a major way. They are also more likely to say that the project had a significant positive impact on the life of their family. Finally, they are much more likely to say that the project has provided them with more family authority, more power over the decisions which most concern them.

Women who have had training are significantly more likely to say that they have less time now for their children or for work in their household and spend more time working. These women are also more likely to say they have the same or less time for their own education or developmental activities, again presumably because they are working more. They are also more likely to say the project is what has led to this change in their time use than women without training. Women without training are significantly more likely to be found in the group whose family income is below the poverty line, although the same percentage (slightly more than a third) are found in the middle or upper income categories. Women who were trained are significantly more likely to say they acquired assets during the project period, while eighty-eight per cent of the women who were not trained said they had not done so. Only in one dimension did untrained women have superiority: they were significantly more likely than trained women to make their own decision on what work they did and how they did it (there was no difference in regard to other decisions).

Women who had had training were also far more likely than those who had not to perceive high project impacts. Most said the project had increased their income (seventy-one per cent), while almost half (forty-eight per cent) of the untrained women said it did not. Seventy per cent of the trained women felt the project had had a major positive impact on their family while only forty-one per cent of those untrained had this perception. Training was obviously a socializing or even mobilizing experience for many women, as forty-nine per cent of the trained women felt the project had given them more decision-making power, while only eleven per cent of the untrained women drew this conclusion. Similarly, women who had had training (ninety per cent) felt their self-confidence and outlook on their own economic future had been improved by the project, while a much smaller percentage of the untrained women did (fifty-four per cent), although it should be pointed out that this too was a majority. There is, of course, a problem with drawing overall conclusions about 'training' as a strategy since this quantitative test does not distinguish which kinds, or amounts, of training the women received nor what their initial skill level was. It does suggest, however, that training – the process of transmitting new skills and outlooks accompanying them – is an essential component in changing attitudes.

In projects where marketing was heavily emphasized, women were much more likely to say that the project had given them a higher income and improved their life than in projects where marketing had not been stressed. They also were far more likely to say the project had had a major impact on their family and that it had given them more decision-making power in their family. Again, we are reporting on the perceptions of the participants and can not control for other project factors but the relationship exists and is significantly different from that observed in projects without a marketing emphasis.

It is noteworthy that three important components do not emerge as having significant impacts on women's lives or their perception of the success of the project: mobilization, credit or financing, and the introduction of a new technology. This does not mean, of course, that these components are not important. In the first place, the variable used for mobilization results in a grouping so that all women participants in each country case fall into one group. This disguises or eliminates differences which might emerge had we information on the degree of mobilization

Table 10.4 Orthogonal transformation solution – varimax

	Factor 1 Income impact	Factor 2 Mobiliza-tion	Factor 3 Decision	Factor 4 Time use	Factor 5 Outlook	Factor 6 Time change economic activity
Project impact on her life	835*					
Project impact on her income	778*					
Project impact on her family	805*					
Membership in a co-operative		.734*				
Ownership assets		-.587*				
Acquire assets during project		-.669*				
Family income		.844*				
Decision on use of income			.859*			
Decision on her work			.9 *			
Change in time spent on rest				.515*		
Change in time spent on own education				.833*		
Project impact on decision-making power					.763*	
Project impact on outlook					.801*	
Project impact on time use						.544*
Impact on economic activities						.832*

* Factor scores present only factor loads of .5 or higher. The negative signs in Factor 2 indicate that 'ownership and acquisition' were coded low to high but 'income' high to low.

each woman received throughout the entire seven country samples. Secondly, the country where mobilization was most emphasized was Tanzania, which is one of the youngest projects, followed by Ghana, the next youngest. It is possible that this time factor outweighs any impacts that mobilization might have in the long run. Thus, too, access to credit (financing) and the introduction of a new technology should be very important, but in our study they may be outweighed by other factors. Until we analyse relative impacts, however, we can not assess their importance.

To know which variables have the largest impact, it is necessary to study the two groups of variables. First, we eliminated three variables: diet, decision on family diet and children's education, none of which related significantly to any other variables in initial tests, probably because these matters were more culturally defined than decisions on income and work. Overall there were seventeen 'impact' variables which we then factor analyzed using a varimax rotation to see if a pattern of factors emerges where certain variables are closely and logically linked together. Our factor analysis yields six factors. The first factor is the 'income impact': it appears that perceptions of improvements to one's own life and that of the family are closely and significantly related to believing that the project increased the participant women's income. The second factor is a 'mobilization and income' factor which is in itself very noteworthy. Belonging to a co-operative (either a prerequisite or a project action in many of our cases) relates most closely to higher family income and the ownership of assets *and* to the acquisition of such assets during the project period. The third factor is a 'decision-making' factor, where the ability to decide on the use of her own revenue is closely related to the decision to decide on whether and how she does her work. 'Time use' forms a fourth factor, as time spent on leisure and rest and time spent on one's own education and self-development are significantly related. The fifth factor, 'outlook', is very interesting as it shows that having decision-making power is closely related to feeling better about oneself and one's future. The last factor, 'time change and economic activities', shows a relationship between believing the project has changed the participant's overall use of time and the change in the overall pattern of economic activities. The factor table above illustrates these relationships.

Taking the mean factor score of the six factors, we now have reduced the analysis to six dependent variables. It is notable here that preliminary tests again showed that one of our independent variables, 'mobilization,'[2] 'did not have any significant relationship with any of the dependent variables constructed from our factor analysis. Mobilization was therefore eliminated from our further tests. The remaining variables showed some degree of intercorrelation as the matrix below illustrates, but not enough to cause concern for severe multicollinearity.

We therefore constructed a series of equations in which the intercorrelated variables were not included simultaneously and ones where they were included to check the effects of these interrelationships. We explored thirty-six models with different combinations of independent and control

---

[2] Mobilization was derived from the project documents and was a scale: high to low: Tanzania, Ghana, Honduras, India, Bangladesh, Peru, Thailand.

Table 10.5 Correlation Matrix*

	Credit	Technology	Training	Age	Education	Time	Market	Level of development
Credit								
Technology	.804							
Training	.499	.599						
Age								
Education								
Time								
Market					-.554			.669
Level of development								

*Only relationships at .5 or above are considered significantly interrelated

variables. In six of the models, the group of independent variables explained a large portion of the variance of the independent variables.[3] We explored all six models and eventually chose the one with the greatest explanatory power. This was the model in which all eight independent and control variables were used. Here there were some significant degrees of interrelationship among technology, training and credit (see the correlation table above). Yet the degrees of interrelationship between these three variables did not cause concern for general overall multicollinearity. In addition, the explanatory power of the inclusive model was higher and there is also an obvious logic to the interrelationship. Where a new or improved technology was introduced, training was provided and credit was often made available to the women. Although not universally connected in this fashion, this was a dominant pattern in our study.

Model I

Y (Income) = f(Credit, Technology, Training, Age, Education, Time, Market, Level of Development)

$$R = .685; F = 27.993_{(p < .001)}$$

Looking at this model in terms of all six factors, we found two relationships to be most significant and interesting. Overall, forty-seven per cent of the total variation in income impact (indicating the relationship between believing the project increased her income and an improvement in her and her family's well-being) was explained. Introduction of a new or improved technology, provision of credit, provision of training and emphasis on marketing were all significantly related to the income impact. At a slightly

---

[3] In our study the proportion of variance explained was much greater than what most social science studies could predict normally because of the complexity of social and economic relationships

lower level of probability, the level of education a women had received also contributed significantly to income impact. Thus we find that the more likely a woman is to report having been introduced to a new technology and trained in it and to have had access to credit, the more she perceives project impact. Having been in a project which stressed and supported marketing also had this effect. The level of development[4] of the country also significantly related to a woman's perception of project impact but the relationship is inverse. That is, the lower the level of development of the country the more a woman perceives that the project as having an impact on her income and therefore on herself and her family.[5] The length of the project also affects perception of impact, the longer the project, the more a woman will be likely to discern impact. Educated women also appear to be more likely to discern project impact. Age, however, has no bearing; neither older nor younger women are any more likely to believe the project affected them or their families.

This finding in itself is a vindication of those projects for women which stress the introduction of a technology and support to it (through training and technology) combined with an emphasis on marketing. The question remains, however, which of these factors are most important in determining the variance. This is a particularly important concern since the age of the project, the level of education *and* the level of development of the country were also significantly related in the data. If the latter three factors were most important, then whatever project strategy is adopted, it may be overridden by these 'circumstance' factors. We looked, therefore at the Beta weights of the different significantly related variables included in this model.

Model I: Regression Summary

Variable	Coefficient	Std Err.	Std. Coeff.	t-Value	Probability	Standard Beta
INTERCEPT	1.126					
Credit	-.421	.114	-.302	3.687	.0003	.36
Technology	.574	.11	.504	5.239	.0001	.47
Training	.225	.072	.199	3.113	.0021	.12
Age	-.004	.021	-.011	.214	.8308	.006
Education	-.053	.027	-.127	1.959	.0512	.011
Market	-.104	.045	-.215	2.331	.0206	.035
Level Dev.	-.085	.028	-.177	3.081	.0023	.018
Time	-.102	.049	-.204	2.077	.0388	.037

---

[4] Level of Development was derived from the per capita annual income of the seven countries (World Bank, 1994, 162)
[5] As Eric Hyman points out this result may reflect percentage change rather than the absolute amount of income gain which would explain why women in the lowest income countries perceived the larger income impacts.

From this analysis, it appears that our project components do have an important impact beyond circumstantial factors such as the level of education of women or the degree to which a country is economically developed. Nor is age of the project particularly powerful in its determinance, explaining only a tiny fraction of the variance in income impact. The introduction of a new technology (Beta weight .47) is first of all important and, given the inter-correlation with credit (the second most important – Beta weight .36) and training (Beta weight .12), we find these explaining women's reactions to a significant extent. Emphasis on marketing although significantly related, explains little of the variation in income impact in contrast.

Significance of quantitative analysis findings

These findings demonstrate a number of interesting points. The major one is that technology transfer is effective in producing positive developmental impacts such as improved outlook on self, on own economic future and in greater decision-making power as long as it is accompanied by training in technology use, available credit or financing and close attention to marketing. Technology transfer on its own, or only including training in its use, does not seem to produce the same results. Nor does credit availability or any other single factor emerge as salient on its own without other combinations. Thus, 'simple' projects using only one (or at most two) intervention strategies are clearly less likely to have as widespread developmental impacts, just as we hypothesized at the beginning.

The operation of this process is through an intermediary or intervening variable – increase in income. That is, technology transfer together with the associated strategies just mentioned does not produce these impacts unless it has led to an improvement in this all-important dimension. Only when participants see that their income has increased do they acquire an improved outlook on their own future, a better lifestyle, more decision-making power etc. Obvious as this may seem, what is demonstrated is important: the direct impact of the technology 'package' described above, filtered through its production of increased income, produces many of the developmental goals identified at the outset of this study.

A number of other independent variables which are not usually considered to be project strategies appear to be important in these results. One of these is length of the project. Several of the tests indicated that the longer a project, the more impacts are registered by the participants. Another finding is the positive influence of education and the reverse relationship with level of development. More educated women more easily perceive the effects on their lives of participating in projects, while women in less-developed (poorer) countries are more likely to register more substantial impacts because, starting from a lower income base as they do, the percentage of the increase is likely to be higher. Both of these findings are evidence of the importance of contextual factors as hypothesized at the outset. They are important primarily because they might influence an evaluation of one project's effectiveness relative to another. These tests show that these types of variables have to be taken into account in project planning. They also show that these factors can be held constant and, when this is done, certain project components do emerge as having specific development results in a wide variety of circumstances and economic and social conditions.

Conclusion

Overall this study has shown through individual cases, and by quantitative comparative analysis, that projects supporting small enterprises do empower women in various ways. Because of an increase in income, especially where this had lasted over a relatively long time, women who participated felt better about themselves. They were generally more optimistic about their futures and they frequently felt that their family position had improved. They had become more important in the family. Their husbands and relatives consulted them. They were able to contribute to family expenses as they had not done before. Although, given the grinding poverty in the regions where they lived, their basic family expense remained food, most felt they could now buy better food. Many of them also had more decision-making power in certain areas such as over what the family ate or how they used their own income. Thus, even in the most desperate situations, changes can occur. Moreover, there is every indication that the changes in outlook and expectation experienced by these women can be passed on to their daughters and may even influence the lives of other women in nearby areas.

The comparative findings do cast more light on some of the assumptions made and hypotheses advanced when this study was designed, but not on all of them. The most complex and impenetrable of these is the role of mobilization, that is, of efforts to actively teach women ways in which they can control their lives and their work and, in the process, raise their own expectations, through group actions. We hypothesized that mobilization had an important effect on women that went beyond or at least enhanced the impacts of other strategies employed in a project. In our quantitative comparison, however, we were not able to demonstrate that mobilization was as important as other project components (such as the introduction of a new technology, training in its use, available credit and help in marketing the product) in producing the positive changes women experienced in their family responsibility, overall status and perception of self-worth. This was because changes in outlook seem to revolve around change in income which in itself seems in these projects to most often result (or be perceived to result) from these combined strategies (technology, training, credit and marketing). Yet there are other intriguing indications which the type of quantitative analysis employed here can not capture.

Disentangling this is important. Some evaluators of SME projects for women have directly focused on past projects which stressed mobilization, to suggest that they tended not to be successful because of their multiple and sometimes contradictory objectives. Mayra Buvinic of the International Center for Research on Women writes, for example, that these projects attempted to achieve too many goals simultaneously, such as integrating women into market production, reducing gender inequalities and/or providing welfare assistance to poor women. Certain common features appeared (according to her), such as an emphasis on group organization and production, on participatory style, on group 'consciousness raising', on skills development in 'female appropriate' tasks subsidized by credit and with some support in marketing. Buvinic continues that projects of this type tend to fail because 'these ambitious aims cloud performance evaluation measures and serve to justify project continuation despite failures in the projects' economic achievements.' (Buvinic, 1993:297–8). Specific reasons for what she calls their 'dismal performance' are threefold: trying to achieve multiple objectives makes project implementation difficult, participatory management is not conducive to economic success, and skills imparted to the participants are often inappropriate and result in 'products which have no or low market demand' (Buvinic, 1993:298). Do our results, then, seem to support this negative assessment of mobilization as a strategy, or perhaps suggest more moderately that mobilization is not needed as desired changes in attitude and action will result simply from economic success?

In fact, the findings here suggest a conclusion which does not support Buvinic's assessments despite the absence of mobilization as a major determinant in improvements in participant outlook in our quantitative comparative analysis. Our results do indicate that changes in outlook and behaviour can result directly from improvements in income, but there are indications as well that mobilization as a strategy has its own importance. Consider first the Tanzanian project. This project has only recently started when this research was undertaken and, at the time and owing to problems in communications, specific village project planning and assessment, and funding, had begun slowly with only a few of the planned village women's groups able to fully implement a plan. Even in this first phase, however, when income improvements have not been experienced by most of the women, what we find is a level of enthusiasm and an optimism which carries the women onward to try to overcome the obstacles they encounter in the very difficult economic situation they face. The difference between them and the majority of the women in Peru is striking. Discouraged and disillusioned, the Peruvian women (except in the one mobilized region!) see nothing to look forward to and no reason to continue with their work on the new technology. In Tanzania, mistakes have been made in village project choice (in some cases because the women and their extension advisers were not realistic about what they could expect to do and how they could sell the product if they succeeded in doing it), but where this type of error did not occur, or was corrected, the women worked together, finding strength and support in their group and its group activities. They expect to overcome their economic problems and they have the will to do so at least in part because of their group mobilization.

Similarly, the women's enthusiasm for their groups and for the support they receive from them as women and in their personal lives rings through in Honduras and in Ghana (in the latter, again, shortness of project experience makes final assessment of the impacts of the emphasis on women's group training impossible). Indications in both cases are that this type of mobilization has been important to the women. Perhaps Buvinic's assessment should be turned around. Clearly a project to support small and medium enterprises (SMEs) for women must be based on effective economic planning: the inputs must be available, the technology cost effective, the skills level (either existent or transmitted) appropriate, the organization intelligible and efficient, the market demand high. This is a necessary condition, a *sine qua non*, but it may not be enough. Mobilization as a strategy provides something more which is a framework for future self reliance. In Bangladesh, for example, women may have experienced the most profound changes in their lives exhibited in this study, perhaps because they started at a lower point than almost any other of our cases – no jobs, no independence, grinding familial poverty and despair. Their project did not mobilize them as a group or emphasize empowerment. The running of the project is dependent on its professional staff who are primarily male. Women process as they were trained to do and the output is marketed for them. They could not take over the processing centre, nor, if it failed, could they organize something to replace it. Their lives have changed for the better; their status has improved in the process, but they could not take responsibility for maintaining or expanding their situation.

The Indian case may illustrate even more about mobilization. In the absence of being mobilized, most of the 500 women project participants (and in all probability the additional 2500 women affected by TADD's sericulture initiative) have achieved not only a marked improvement in income but also a striking degree of change in their family position and outlook. Only a handful of women in six villages have experienced the mobilization efforts of ASTHA yet the others have also enjoyed profound changes, changes which were probably *under*-valued in initial assessments of project impact. What the women in the Mahila Mandals which ASTHA has trained have, which the others do not, seems to be an orientation to their own capacity for action and change both in and outside their families.

This is reflected in the things that they now do. Women in these Mahila Mandals have inserted themselves into the local political situation. This is in striking contrast to the silent and passive role accepted by Bhil women previously. Mobilization, thus, has helped them act as a group in the community. It has helped them to see their own worth in contradistinction to traditional views of women in their society, and has thus given them a power for action quite unlikely for women who are not socialized into group action. The India case suggests therefore that mobilization may be the strategy which is necessary if women are to take a significant part in, and responsibility for, changing their future. This finding parallels the findings of other scholars who have argued that multisectoral programmes (including mobilization) are preferable 'provided these were conceptualized and implemented in a staged and coherent manner' (Beets, Neggers and Wils, 1988:55).

Among other things, these studies indicate that projects which emphasize group solidarity, training and awareness as well as economic objectives *do* empower the recipients and may lead to major changes in their political and social patterns which ultimately also benefit them economically. Thus BRAC, in contrast to the Surjosnato project in Bangladesh, does result in changed patterns of community involvement with increased responsibility and authority resulting for the mobilized BRAC women (Wils, 1993). Our study, then, appears to support the importance of mobilization as a strategy although, to prove its impacts quantitatively, the study would have had to add measures of community activities and responsibility and project self-reliance.

Buvinic's comments denigrating projects stressing mobilization are placed in the context of a criticism (for the reasons cited above) of multisectoral as opposed to monosectoral projects or of complex (many-factor) as opposed to simple (single-factor) SME projects. The question raised here in this regard was whether simple projects could or would have the same developmental impacts as the complex alternatives (and, of course, which combination of strategies among the complex projects was most effective). We hypothesized that complex projects would have wider and deeper developmental impacts. Only two of the projects considered here could be termed 'simple' or monosectoral. In one (Peru), the emphasis was purely on the transmission of new or adapted technologies. In the other (Thailand), the only intervention was the availability of venture capital (with some business and marketing assistance). In neither of these cases were developmental impacts observed except in the one region of Peru where mobilization as well as business and marketing assistance had been provided for the women's groups who participated. Thus indications from this study are that developmental impacts may well be less from so-called simple or monosectoral projects.

This parallels a later finding from an evaluation of a credit project in Senegal, also monosectoral. Here, only credit was made available. This credit was valuable for the businesses concerned, but few developmental impacts could be observed for the majority of the participants, although these were greater among the women than for the men because the proffered credit was a major (as opposed to just one of many) source of funds for the women (Vengroff and Creevey, 1994). This point should not be taken to mean, however, that such simple projects have no purpose. Where funds are scarce and the economy stagnant, breaking down one of the barriers to productivity and growth facing entrepreneurs has to be a significant achievement. This addition to their resources will, however, naturally have less of an impact on their lives because the change it engenders in their economic activities or their specific business or business practices is much smaller than in the more complex cases.

Jeffrey Ashe suggests that simple projects work best in situations where there is a growing and expanding economy, and less well in situations where the economy – and the market – is contracting (Ashe, 1985:28). This study provides an elementary support for this contention – the Thailand project had the planned result (money was generated and reinvested) – in a growing and expanding economic situation, while the Peruvian project produced many new technologies but few successful enterprises in a economy paralysed by civil strife, disease and economic stagnation. Whether or not a complex project would have worked in Peru can not be proved here but projects in zones similarly devastated by civil strife and facing major economic problems, such as those in Honduras and Guatemala, were more successful and the developmental impacts were, of course, greater, as pointed out above.

Among the desired impacts was an improvement in the women's quality of life, including in its scope an easier and more restful living pattern for the poor women who currently labour from dawn to dusk. One finding which has been repeatedly discussed in the individual cases is that projects which successfully generated income for women usually did not lighten their workload. If anything, the majority of these schemes resulted in women working harder than they had. The one case where this appeared not to be so was in Ghana and for a specific reason. In Ghana, the new technology was introduced where women were already producing shea butter using traditional methods. The new technologies drastically reduced the preparation time, and thus the work time of the participants was cut.

In virtually all of the other cases, women learned how to produce something which they had not produced before. The technologies were 'appropriate', that is, they were low cost and simple, but the new production was additional work. Women continued to do what they already did in terms of domestic work and work in subsistence agriculture *and* they began to produce a new or modified product. The women participants themselves did not fault this result. They accepted it and even embraced it for the benefits they perceived. Although increased income was clearly more important to them than rest, in terms of project preparation, this study suggests that planners should be aware of this probable impact accompanied by lessened time for the children and for the household, although women continued to be the responsible persons in these areas. For most women, then, their quality of life was improved in terms of their better outlook on themselves and their future, and the benefits they and their family reaped from an increased income, but in regard to their ease of living there was a deterioration, not an improvement, for most.

Another controversial area was our definition of independence and our search for increased independence as a measure of a developmental impact. Our findings suggest that in countries where women had the least traditional access to economic activities outside of domestic work, such as Bangladesh and India, women gained directly in terms of perceived independence measured by increased authority over their income and work and within other family decision-making arenas. But, in countries where women traditionally carried on economic activities outside the home and had a marked degree of independence, such as Ghana, the result of the project seemed almost to be a loss in independence as men began to take part in decisions about the allocation of the women's income. This was true in the similar study of developmental impacts carried out in Senegal in 1994 (Vengroff and Creevey, 1994). Yet, quite inconsistently, the women in these projects felt they had gained in authority and improved their family position.

On the one hand, it is possible to interpret this negatively, as some observers have in similar situations. They claim that, as soon as women begin producing something of value, men want to take it over and get control of the profits for themselves, thus ultimately lessening women's choices and their authority as they increasingly become dependent on their men (Thiam, 1986:76). On the other hand, other interpretations are possible and may be more useful in this evaluation. The Ghanaian women did not observe that they had lost independence or authority. In fact they believed the project has made them feel better about themselves and about their future. They cited things which they were able to do for the family with their money which they could not have done before. They seemed to see themselves as *more* important to their families now because they could give more. As women in other projects pointed out, 'my husband and my relatives ask my opinion now as they did not before'. In other words, they have gained status.

Their enhanced occupational role has led to an enhanced domestic role although their actual independence is somewhat curtailed. Men's attention to their work and their income is in itself a sign that what the woman does has become more important. The men may aver that the women's income has not changed the family economic situation. Particularly in cultures where men are expected to provide for the family and women are not, this will be an automatic reaction. But a clear sign of the increased value of what the woman does is that the man will now want to discuss what she is doing and how she will spend her money. It may still be worthwhile to measure (as far as possible) changes in authority over decisions. First of all, in societies where women had no scope for economic decision-making, an acceptable developmental goal may be an increase in their authority and independence. In other societies where women already are fully involved in the public sector it still may be important to measure changes in authority and independence because at some point, if men do take over all the tasks and responsibilities and decisions of the SME, women may become more dependent than they were and probably will lose status as well.

Economic dependence is not necessarily seen by the women themselves as bad as long as they get the material things and lifestyle they want for their children and themselves (Oppong and Abu, 1985). A decrease in family status, however, measured by *less* responsibility and *more* treatment as an inferior not worth consulting, would probably be a negative even to women traditionally raised in a strict conservative society. In any case, the overall (Western) judgment that women should gain in authority over their lives and work may still determine the ultimate scale of developmental impact, but more refined and careful allowance for the probable lags and switchbacks in the process is necessary in designing an evaluation.

Equally problematic is assessment of the importance of 'trickle-down' impacts from men's projects to the women in their family. This study is based on the premise that projects should work directly with women in order to achieve the maximum impact on women's lives and work. Nothing in the findings here contradicts this assumption but another refinement does seem to be suggested by our cases. In Guatemala, project managers contend that they could not have started trying to directly involve women in the Wool Production and Processing Project because of existing traditional attitudes. They feared their overall project might have been undercut by the opposition such a move might have engendered. Whether or not this is the case, they were able in a later phase of the project to begin to work with women's groups.

Beyond this, their programme for the men had many discernible positive impacts on the women in their families, including not only improvements in lifestyle, but also a kind of mobilization, an improvement in self-image, and an increased demand for women's activities to be carried out further. In Honduras women were initially targeted, although at the outset the project was primarily for men. What is interesting in this context is the step process which inadvertently occurred here as well. Women became more important and their role changed as other factors changed in the situation. It is possible that a kind of stage process could be built in to take advantage of the changes in outlook that women experience as their men are drawn into a programme. Working with women from the beginning still seems desirable if the objective of the agency planning to intervene is an improvement in their situation. But for others, perhaps local government agencies, trying in a more general sense to establish SMEs, or support and expand existing ones, recognizing and utilizing the trickle-down in step-process planning can expand their results and, in all probability, improve their overall achievement of project objectives.

Another criterion for the evaluation of SME projects mentioned at the outset is the combined importance of project length and project sustainability/replicability. This study has dramatically illustrated the importance of project length in the achievement of developmental goals. Projects which have gone through many hardships, had limited funding and major problems in implementation, such the Honduras or Bangladesh projects, nonetheless had major impacts. Comparative quantitative analysis suggests that the length of the experience is an important factor in the degree to which developmental changes result. Obvious though this may seem, it is still passed over by some funding agencies which insist on relatively short project spans because of their own budgeting cycle. Yet full impacts, and even the nature of those impacts which are actually occurring, may not be visible until the project has been in existence ten years or more. Of course the project may be of shorter duration than this, but, assuming it sets up a viable enterprise or introduces a new technology to such an enterprise, it will still be some time before it is possible to see what the results really are. This argues for scheduling evaluations long after the project end to discern what has happened to the women involved and what in the project caused this to happen.

But which projects, with which characteristics or particular combination of strategies, were most likely to become long-term sustainable projects reaching large numbers of women in any case? In our study the project which reached the most women was the Sericulture Project in India. It was not only the programme which has had the widest impact but also it appears to have a very good likelihood of being sustained and replicated for the foreseeable future. Our study suggests that one very important factor is the involvement of the state government. This ensured a large, region-wide staff to promulgate the project (and to wave aside normal governmental red tape) and able to negotiate with donors and for markets with national government backing.

There is still the question of what happens if and when the enterprise has disappeared, swallowed perhaps by competition from another enterprise with more sophisticated technology, management, organization and marketing? How then can the project be evaluated? Is it a failure if the enterprise is not able to compete after a relatively short time? What happens to the women who have been in the project when this occurs?

In many of our cases future viability was a serious concern. In Honduras, for example, the cashew nuts did not produce a high enough income to compete with jobs thathad become available to the men in other fields of endeavour as their zone began to open up economically. The cashews produced were not of as high quality as those produced in some other countries (and in other projects in the same country). In the long run, could even the women afford to stay with this enterprise? In Bangladesh, the Coconut Project was being forced to introduce other types of activities, such as growing seedlings and vegetables, as its products were unable to find, and keep, an adequate market while inputs (coconuts) were progressively getting more expensive. In India, the opening of the country's market (to Chinese products amongst many others), and a World Bank-sponsored sericulture project with a higher level of technology, threatened the market for the products of the Rajasthan women. And, finally, in Ghana, the high rate of return of the shea butter produced with the new technology had attracted Ghanaian businessmen who were thinking of establishing a factory to produce shea butter which might well undercut the profits the women now make.

Although we can not know what will happen to these women, it seems that these projects can not be judged failures. In all cases they established enterprises which lasted over relatively long periods of time (or, for Ghana and Tanzania, are likely to last over quite a long period), and had relatively large impacts on the people involved (or, for the latter two, are likely to have such impacts once fully underway). In any case, the whole process of development implies absorption by the smaller, less-efficient industries by larger, more complex industries over time. Those affected have to retrain and engage in new endeavours or become part of the wage-paid labour force. The hope must be that the former participant women will be able to turn their experience to another enterprise or another income-generating opportunity, take the profits and invest in the new endeavour.

As to the key question of this research – which combination of strategies is the best or most effective in achieving an increase in income combined with improved authority and outlook patterns – this study has a peculiar deficiency: lack of project diversity. This is hard to credit given what has been described here. As stated in the first chapter, these were very varied projects in extremely dissimilar economic, political and social situations. Sponsored by international agencies with different backgrounds, philosophies and approaches to their work, the projects promised to be difficult to compare. The entire situation was complicated by the wide range of local NGOs, government agencies or other international agencies through which all of the projects sponsored by our three international agencies worked. Yet, one factor was a constant in all but one of these projects – an emphasis on the introduction of a new or improved technology combined with training in its use. Our finding, through quantitative cross-national and individual case analysis was that the introduction of a new or modified technology, combined with training and credit (and to a lesser extent marketing) had an impact which was measurable by improvements in income, combined with a belief that the project had improved the woman participant's life and that of her family.

In itself this result is interesting, particularly since we can show that technology introduction without credit and marketing assistance is *not* effective in achieving that same end. And, we can also show that observed impacts were not due to exogenous factors, such as women participants' ages, their levels of education, or even simply project length. However, we can not show what happens in projects where the introduction of a new technology was not part of the package offered. Our one case does not give us anything to add to the picture, since the monosectoral projects in our study have proven to have the fewest long-term impacts on their participants. We do not know how projects which emphasize strategies other than technology, such as, to take just one example, introducing a new enterprise using existing skills and domestic technologies, but combined with credit and marketing assistance, compare to our cases.

In the absence of further cases, our conclusions on strategy effectiveness can only be that technology transfer in the right combination is effective in terms of producing increased incomes and associated changes in outlook and improvements in status. This is on one level disappointing, but it should not be understated. It is extremely important to recognize that a *package* of interventions including technology may have broad-ranging developmental impacts, where introducing a new technology *per se* will not.

One other added indication from our study suggests that technology introduction, training, credit and marketing assistance may be most effective where the entire chain of production is taken into account. The Guatemala project can not be directly compared to the other seven because of the gender of its participants, but the Guatemala case does illustrate the value of the sub-sector approach which allows the manipulation of the chain of production and the final market. Such an approach may require a relatively high investment, but promises better sustainability if fully pursued. In this sub-sector context, the optimal use of new and improved technologies is better ensured than where such alterations in method of production face market and input constraints which they may not be able to overcome.

As a final caveat to this research, it is necessary to reiterate that nothing here is definitively established beyond the need for further exploration or testing. We merely have drawn out of admittedly limited quantitative and qualitative data, those trends or relationships which seemed most striking and important. Further work is advised to substantiate some of the suggestions or interpretations which we have advanced. But these findings need to be considered immediately to avoid future errors (costly to the economically marginal SME participants and the donors themselves) as factors necessary in project planning and evaluation. We would argue, based on what has emerged here, that these are principal points:

1) Projects aimed at the poorest women not yet established in sustainable enterprises must be multifaceted to be successful.

2) Projects intending to have an impact on community organization, or to allow women to take full charge of their own businesses, as well as take a more active role in their communities' social and political life, must include mobilization training. This implies the training of women in groups, not only in group organization, management skills etc. but also consciousness raising.

3) Introduction of new or modified technologies should be embedded in a larger package containing (at least) management training, marketing assistance and access to credit.

4) One useful approach may be to refer back to projects formerly only aimed at men and assess what women in their families learned/gained from these and build on this experience with new projects planned specifically for them.

5) Project planning must include an awareness of probable project cost to the participants in terms of increased workload and less time for family care.

6) The extent to which women will experience an increase in their decision-making power as a result of participating in a project supporting SMEs will vary depending on whether the environment already promotes women's independent economic roles. The more it does so, the less likely it is for there to be such a result. In any case, however, an increase in family status (respect and importance) should result.

7) Projects assisting SMEs (which take the above six factors into account), may have significant impacts on the income, assets, time use, decision-making power and outlook of the women who participate in them.

CHAPTER TEN: Comparative results of significant ANOVA, Chi Square, and cross tabulations using Kendall's TAU

ANOVA
1) Sample Type (Participant, Control) and Membership in Co-operative (No, Yes)
   Sum squares = 106.117, Mean difference = .531, p. = .0001
2) Sample Type (Participant, Control) and Perception of Change in Economic Activities (No change, Change)
   Sum squares = 989.785, Mean difference = 1.769, p. = .0001
3) Sample Type (Participant, Control) and Use of Time in the Household (More, Same, Less)
   Sum squares = 287.11, Mean difference = .423, p. = .0001
4) Sample Type (Participant, Control) and Use of Time for Leisure (More, Same, Less)
   Sum squares = 310, Mean difference = .842, p. = .0001
5) Sample Type (Participant, Control) and Use of Time for Self Development (More, Same, less)
   Sum squares = 125.439, Mean difference = .447, p. = .0001
6) Sample Type (Participant, Control) and Impact of Project on Time (No impact, Impact)
   Sum squares = 84.183, Mean Difference = .633, p. = .0001
7) Sample Type (Participant, Control) and Ownership of Assets (Has none, Has few, Has many)
   Sum squares = 103.063, Mean Difference = .122, p. = .0064
8) Sample Type (Participant, Control) and Family Income (High, Middle, Low, Below Subsistence)
   Sum squares = 300.517, Mean Difference = -.215, p. = .0054
9) Sample Type (Participant, Control) and Decision on Work (She decides, Husband decides, She and He decide together)
   Sum squares = 71.515, Mean Difference = -.095, p. = .-191

Cross Tabulations (with Kendall's Tau)

Participants only:
10) Age of Project (Recent, Old) and Impact of Project on Participant's Life (No Impact, Positive Impact)
    Kendall's Tau -b= .52468
11) Age of Project (Recent, Old) and Perception that Income Increased because of Project (No, Yes)
    Kendall's Tau -b = .52168
12) Age of Project (Recent, Old) and Perception that Project Positively Affected her Family (No, Yes)
    Kendall's Tau -b= .67847
13) Age of Project (Recent, Old) and Perception that Project Helped her to Get Greater Authority (No, Yes)
    Kendall's Tau -b = .56691
14) Market Emphasis (No, Yes) and Impact of Project on Participant's Life (No Impact, Positive Impact)
    Kendall's Tau -b = .68755
15) Market Emphasis (No, Yes) and Perception that Income Increased because of Project (No, Yes)
    Kendall's Tau -b = .61643
16) Market Emphasis (No, Yes) and Perception that Project Positively Affected her Family (No, Yes)
    Kendall's Tau -b = .54291
17) Market Emphasis (No, Yes) and Perception that Project Positively Affected her Decision Making Power (No, Yes)
    Kendall's Tau -b = .37811

Chi Square

Participants only:
18) Provision of Training by Project (No, Yes) and Change in Use of Time for Leisure (More, Same, Less)
    Chi sq. = 30.938, Cramer's V = .348, p. = .0001
19) Project Provision of Training (No, yes) and Change in Use of Time for Self-Education (More, Same, Less)
    Chi sq. = 36.543, Cramer's V = .456, p. = .0001
20) Project Provision of Training (No, Yes) and Perception that Project Had an Impact on Overall Time Use (No, Yes)
    Chi sq. = 25.275, Phi = .318, p. = .0001

21) Project Provision of Training (No, Yes) and Family Income (High, Middle, Low, Below Subsistence)
Chi sq. = 9.702, Cramer's V = .195, p. = .0213

22) Project Provision of Training (No, Yes) and Acquisition of Assets During Project (None, Few, Many)
Chi sq. = 16.418, Cramer's V = .259, p. = .0003

23) Project Provision of Training (No, Yes) and Decision on Whether and How She Works (She Decides or Decides in Consultation, Husband Decides)
Chi sq. = 5.783, Phi = .152, p. = .0162

25) Project Provision of Training (No, Yes) and Perception that Project has Increased Her Income (No, Yes)
Chi sq. = 12.72, Phi = .225, p. = .0004

26) Project Provision of Training (No, Yes) and Perception that Project has had a Positive Impact on her Family (No, Yes)
Chi sq. = 20.2, Phi = .28, p. = .0001

27) Project Provision of Training (No, Yes) and Perception that Project has given her more Authority (no, Yes)
Chi sq. = 32.35, Phi = .365, p. = .0001

28) Project Provision of Training (No, Yes) and Perception that Project has given her more Self-Confidence and a better Outlook on her Economic Future (No, Yes)
Chi sq. = 41.535, Phi = .409, p. = .0001

Bibliography

Arias, Maria E. and John Ickas (1985) 'Peru: Banco Industrial del Peru; Credit for the Development of Rural Enterprise,' in Catherine Overholt, Mary B. Anderson, Kathleen Cloud and James E. Austin, *Gender Roles in Development Projects: A Case Book*. West Hartford: Kumarian Press.
Ashe, Jeffrey (1985) *The Pisces II Experience; Local Efforts in Micro-Enterprise Development*. Washington: Agency for International Development, (April).
ATI (1991) 'Report on the Feasibility of Establishing A Medium-Scale Cashew Processing Enterprise in Choluteca, Honduras'. Washington. (June).
ATI (1993) 'A Window on ATI,' Washington: Appropriate Technology International.
Bagachwa, M.S.D. (1993) 'Impact of Adjustment Policies on the Small-Scale Enterprise Sector in Tanzania,' in A.H.J. Helmsing and Th. Kolstee, eds., *Small Enterprises and Changing Policies: Structural Adjustment, Financial Policy and Assistance Programmes in Africa*. London: IT Publications.
Bakht, Anjana (n.d.) 'Dawn,' Report on the sericulture project prepared for UNIFEM.
Beets, Nico, Jan Neggers and Frits Wils (1987–1988) 'Big and Still Beautiful; Enquiry into the Efficiency and Effectiveness of Three Big NGOs (BINGOs) in South Asia', The Hague: DGIS/NOVIB (Nov–April).
Begazo, Victor Raul, Lucila Caceres and Francisco Verdera, *Alternative Technologies for Food Processing in Rural Areas of Peru, External Evaluation of March 1992*.
Berenbach, Shari and Diego Guzman (1992) *The Solidarity Group Experience Worldwide*. Washington: ACCION (November).
Berger, Marguerite and Mayra Buvinic, eds. (1989) *Women's Ventures; Assistance to the Informal Sector in Latin America*. West Hartford: Kumarian Press.
Bhatt, Ela (1989) 'Toward Empowerment,' *World Development* XVII (7) (July).
Bourque, Susan C. and Kay Barbara Warren (1981) *Women of the Andes: Patriarchy and Social Change in Two Peruvian Towns*. Ann Arbor: University of Michigan.
Brandler, Natalia (1994) 'Indigenous Women and Strategies for Small Enterprise Development,' Storrs: Unpub. Mss. (May).
Bromley, Ray and Chris Gerry, eds. (1979) *Casual Work and Poverty in Third World Cities*. New York: John Wiley and Sons.

Bryceson, Deborah Fahy (1985) 'Women's Proletarianization and the Family Wage in Tanzania,' in Haleh Afshar, ed., *Women, Work and Ideology in the Third World*. London: Tavistock Publications.

Buvinic, Mayra (1989), 'Investing in Poor Women: The Psychology of Donor Support,' *World Development*, XVII (7) (July).

Buvinic, Mayra (1993) 'Promoting Women's Enterprises: What Africa Can Learn from Latin America,' in A.H.J. Helmsing and Th. Kolstee, eds., *Small Enterprises and Changing Policies: Structural Adjustment, Financial Policy and Assistance Programmes in Africa*. London: IT Publications.

Callaway, Barbara and Lucy Creevey (1994) *The Heritage of Islam; Women, Religion and Politics in West Africa*. Boulder: Lynne Rienner Publishers.

Carr, Marilyn (1984) *Blacksmith, Baker, Roofing-sheet Maker . . . ; Employment for Women in the Third World*, London:IT Publications.

Catanzarite, Lisa (1992) 'Gender, Education, and Employment in Central America: Whose Work Counts?' in Nelly Stromquist, ed., *Women and Education in Latin America; Knowledge, Power and Change*. Boulder: Lynne Rienner Publishers.

Chazan, Naomi, Robert Mortimer, John Ravenhill and Donald Rothchild (1992) *Politics and Society in Contemporary Africa*. Boulder: Lynne Rienner Publishers.

Chen, Marty (1987) 'Developing Non-Craft Employment for Women in Bangladesh,' in Ann Leonard, ed., *Seeds: Supporting Women's Work in the Third World*. New York: The Feminist Press.

Cheru, Fantu (1992) 'Structural Adjustment, Primary Resource Trade and Sustainable Development in Sub-Saharan Africa,' *World Development*, Vol. 20 (4).

Chiang Rai Agro Industries (1988) Investment Proposal (December).

Chickering, A. Lawrence and Mohammed Salahdine, eds. (1991) *The Silent Revolution: The Informal Sector in Five Asian and Near-Eastern Countries*. San Francisco: International Center for Economic Growth.

Dak, T.M. (1988) *Women and Work in Indian Society*. New Delhi: Discovery Publishing House.

Dawson, Jonathan (1993) 'Impact of Structural Adjustment on the Small Enterprise Sector: A Comparison of the Ghanaian and Tanzanian Experiences,' in A.H.J. Helmsing and Th. Kolstee, eds., *Small Enterprises and Changing Policies: Structural Adjustment, Financial Policy and Assistance Programmes in Africa*. London: IT Publications.

Delp, Peter *et al.* (1986) *Promoting Appropriate Technological Change in Small-Scale Enterprises: An Evaluation of Appropriate Technology International's Role*. Washington: USAID (November).

Desai, Neera (1988) *A Decade of the Women's Movement in India*. New Delhi: Himalaya Publishing House.

Desai, Neera and Vilohuti Patel (1985) *Indian Women; Change and Challenge in the International Decade 1975–1985*. Bombay: Gibson Publishing Works.

De Wilde, Ton, and Steinje Schreurs with Arlene Richmond (1990), *Opening the Marketplace to Small Enterprise; Where the Magic Ends and Development Begins*. West Hartford: Kumarian Press.

Dhamija, Jasleen (1989) 'Women and Handicrafts: Myth and Reality,' In Ann Leonard, ed., *Seeds: Supporting Women's Work in the Third World.* New York: The Feminist Press.

Donkor, Peter (1991) 'Development of an Improved Shea Butter Processing Technology,' Consultancy Report for GRATIS Project.

Downing, Jeanne (1990) *Gender and the Growth and Dynamics of Microenterprises.* Washington: GEMINI, September.

Drake, Susan and Mark Sullivan, eds. (1993) *Appropriate Technology International l992; Annual Report.* Washington: ATI.

Dulansey, Maryanne and James Austin. (l985) 'Small-Scale Enterprise and Women,' in Catherine Overholt, Mary Anderson, Kathleen Cloud and James Austin, eds. *Gender Roles in Development Projects*, West Hartford: Kumarian Press.

Elmendorf, Mary Lindsay (1972) *Mayan Women and Change.* Cuaderno 81. Cuernavaca: Centro Interculteral de Documentacion.

Europa (1992) *The Europa World Yearbook, l992,* Vol. 1. London: Europa Publications Limited.

Fine Dried Fruits International (1991–1992) 'Honduran Cashew Fruit Drying: Work Performed by Rusty Brown.'

Fortmann, Louise (1982) 'Women's Work in a Communal Setting: The Tanzanian Policy of Ujamaa,' in Edna G. Bay, ed., *Women and Work in Africa.* Boulder: Westview Press.

French, Jerome (1988) 'Sustainability Study,' Internal Evaluation Document for the Office of Rural Development, Washington: USAID.

Ghandi, Nandita and Nandita Shah (1992) *The Issue at Stake: The Theory and Practice in the Contemporary Women's Movement in India.* New Bakersville: IPP Publications.

Gopinath, C. and A.H. Kalro (1985), 'India: The Gujarat Medium Irrigation Project,' in Catherine Overholt, Mary Anderson, Kathleen Cloud and James Austin, eds. *Gender Roles in Development Projects*, West Hartford: Kumarian Press.

Government of Honduras (GOH) (1988) *Caracteristicas Generales, Educativas y Economicos por Dept.* Tomo 1: *Censo Nacional de Vivienda 1988.* Tegucigalpa.

GRATIS (1988) *Annual Review.*

Grown, C.A. and J. Sebstad (1989) 'Introduction: Towards Wider Perspectives on Women's Employment,' *World Development* XVII (7) (July).

Gupta, S.K. (1993) Fax to L. Creevey (Oct 18).

Hyman, Eric (1992) 'Enhancing the Productivity of Small Firms,' Lecture and Discussion Notes, Kingston, Jamaica (Sept 22–24).

Hyman, Eric, S.K. Gupta and Ashvin Dayal (1993) 'Comments on the Pickled Ginger Enterprise' (Aug 3).

Heyzer, Noeleen (1988) 'Asian Women Wage-Earners: Their Situation and Possibilities for Donor Intervention,' *World Development* Vol 17 (7) (July).

Hook, Margaret (1993) *Guatemalan Women Speak.* Washington: EPICA.

Institute for Socio-Economic Research (IIES) (1990) 'Informe, Proyectos de Procesamiento de Maranon,' Honduras.

Intermediate Technology Development Group (ITDG) (1992) *Annual Report 1991/1992*. Rugby: ITDG.
Intermediate Technology Development Group (ITDG) (n.d.) 'Shea Butter Production, Tamale ITTU,' PN 86.2542–03.100.
Intermediate Technology Development Group (ITDG) (1988) 'Surjosnato Coconut Project Bangladesh; Technical and Economic Evaluation,' (June).
International Monetary Fund (IMF) (1992) *International Financial Statistics Yearbook 1992*. Washington: IMF.
Jahan, Roushan (1991) 'Women's Movement in Bangladesh; Concerns and Challenges,' in Neuma Aguiar and Thai Corral, eds. *Alternatives: Women's Visions and Movements*. Rio de Janeiro: Editora Des Tempos.
Leonard, Ann, ed. (1989) *Seeds: Supporting Women's Work in the Third World*. New York: The Feminist Press.
Liedholm, Carl and Donald Mead (1987) 'Small Scale Enterprises in Developing Countries: Empirical Evidence and Policy Implications,' Michigan State University, International Development Paper No. 9, East Lansing: Michigan State University.
Little, Ian, Dipak Mazumder and John Page (1987) *Small Manufacturing Enterprises: A Comparison of India and Other Economies*, Washington: World Bank Research Publication.
Loarca, Alfonse (1990) 'Evaluacion Programma Momostenanga: Washington: ATI Guatemala documents (February).
Lobo, Adonay (1993) 'Informe des Dept. des Capacitacion de Asociacion des Proyectos de Pueblo,' (Jan). Tegucigalpa: GOH.
Lovell, Catherine (1989) 'Case 2: BRAC (A); How to Define "Self-Supporting," in the BRAC Rural Credit Program,' in Charles K. Mann, Merilee S. Grindle and Parker Shipton, eds., *Seeking Solutions: Framework and Cases for Small Enterprise Development*. West Hartford: Kumarian Press.
McClintock, Cynthia (1989) 'Peru: Precarious Regimes, Authoritarian and Democratic,' in Larry Diamond, Juan J. Linz and Seymour Martin Lipset, eds., *Democracy in Developing Countries, Volume Four: Latin America*. Boulder: Lynne Rienner Publishers.
McKean, Cressida and Annette Binnendjik (1988) 'AID's Small-Enterprise and Microenterprise Projects; Background and Current Issues,' AID Occasional Paper No. 18, Washington: USAID.
Mann, Charles K., Merilee S. Grindle and Parker Shipton, eds. (1989) *Seeking Solutions: Framework and Cases for Small Enterprise Development*. West Hartford: Kumarian Press.
Marina Delgado, Luz (1992) 'Diagnostico de la Situacion de la Mujer en la Zona des los Chuhumatanes, Huehuetenango,' Quetzaltenango (July 30).
Mbilinji, Marjorie (1989) 'Women as Peasant and Casual Labor and the Development Crisis in Tanzania,' in Jane Parpart, ed., *Women and Development in Africa: Comparative Perspectives*. Lanham: University Press of America.
Mbughuni, Patricia, Nancy A. Masumba and Japeth M.M. Ndaro (1989) 'Report of the Internal Evaluation of the Women's Appropriate Food

Technology Project (URT/87/WO1),' Prepared for the Department of Community Development and UNIFEM (October).

Mennonite Central Committee (MCC) (1992) *Job Creation Program; Annual Report 1991–1992*. Bangladesh.

Nanjundan, Subrahmanyan (1989) 'Should SSE Policy be an Integral Part of Overall Development Policy?' Paper delivered at the Workshop on Small-Scale Enterprise Development: In Search of New Dutch Approaches, The Hague.

Oppong, Christine and Katherine Abu (1985) *A Handbook for Data Collection and Analysis of Seven Roles and Statuses of Women*. Geneva: International Labour Office.

Osei, Barfour, Amoah Baah, Nuakoh, Kwadwo Tutu and Nii Kwaku Sowa (1993) 'Impact of Structural Adjustment on Small-Scale Enterprises in Ghana,' in A.H.J. Helmsing and Th. Kolstee, eds., *Small Enterprises and Changing Policies: Structural Adjustment, Financial Policy and Assistance Programmes in Africa*. London: IT Publications.

Otero, Maria (1989) *A Question of Impact; Solidarity Group Programs and Their Approach to Evaluation*. Tegucigalpa: PACT.

Palmer, David Scott, ed. (1994). *The Shining Path of Peru*. Second Edition. New York: St Martin's Press.

Pongsapich, Amara (1991) 'Women's Social Protest in Thailand,' in Neuma Aguilar and Thais Corral, eds. *Development Alternatives for a New Era*. Rio de Janeiro: Editora Rosa dos Tempos.

Pueblo-to-People (P-P) (1993) *Project Documents for the Honduran Cashew Nut Project*. Washington DC.

Reichmann, Rebecca (1989) 'Women's Participation in Two PVO Credit Programs,' in Marguerite Berger and Mayra Buvinic, eds., *Women's Ventures; Assistance to the Informal Sector in Latin America*. West Hartford: Kumarian Press.

Revere, Elsbeth (1990) 'Guatemala; A Thriving Undertaking, Private and Public Programs Serving the Informal Sector,' in Katherine Stearns and Maria Otero eds., *The Critical Connection; Governments, Private Institutions, and the Informal Sector in Latin America*. Washington: ACCION (August).

Rios Varillas, Walter and Gonzalo la Cruz (1981) 'Alternative Technologies for Food Processing in Rural Areas of Peru', ITDG Proposal (December).

Roberts, Penelope (1987) 'The State and the Regulation of Marriage: Sefwi Wiawso (Ghana), 1900–1940,' in Haleh Afshar, ed., *Women, State and Ideology: Studies from Africa and Asia*. Albany: State University of New York Press.

Robertson, Claire C. (1984) *Sharing the Same Bowl: A Socioeconomic History of Women and Class in Accra, Ghana*. Bloomington: Indiana University Press.

Rodriguez-Streeter, Guadelupe (1994) 'Evaluating the Rajasthan Project,' Unpub. Mss. Storrs (May).

De Soto, Hernan (1989) *The Other Path*. New York: Harper and Row Publishers.

Sharma, Kumud (1991) 'Women's Movement in India: Dialectics and Dilemma,' in Neuma Aguiar and Thai Corral, eds. *Dawn: Development Alternatives for Women for a New Era*. Rio de Janeiro: Editora Dos Tempos.
Sjostrom, George (n.d.) 'Traditional Shea Butter Extraction in Northern Ghana,' Gratis Project paper.
Spade, Maria (1989) 'Informe de Progresso.' New York: UNIFEM reports.
Stearns, Katherine and Maria Otero, eds. (1990) *The Critical Connection: Governments, Private Institutions, and the Informal Sector in Latin America*. Washington: ACCION (August).
Steel, William F. and Claudia Campbell (1982) 'Women's Employment and Development; A Conceptual Framework Applied to Ghana', in Edna G. Bay, ed., *Women and Work in Africa*. Boulder: Westview Press.
Stocks, Kathleen and Anthony Stocks (1984) 'Status de la Mujer y Cambio por Aculturacion: Casos del Alto Amazonas,' *Amazonia Peruana*, Vol. V, No. 10.
Tawakley, Kalpana (1992) Memo to Madhu Bala Nath (August 31).
Thiam, Miriama (1986) 'The Role of Women in Rural Development of the Segou Region of Mali,' in Lucy Creevey, ed., *Women Farmers in Africa; Rural Development in Mali and the Sahel*. Syracuse: Syracuse University Press.
Tokman, Viktor E. (1989) 'Policies for a Heterogeneous Informal Sector in Latin America,' *World Development* XVII (7) (July).
Torrens, James (1991) 'Central America: L and the Burning Issue,' *America* 165 (Sept 14).
Traitongyoo, Tavatchai (1989) Letter to Ton de Wilde, Executive Director, ATI.
UNIFEM (n.d.) Reports on the Sericulture Project: IND/88/WO1 A/72/99 'Project Document' and 'Integrated Development of Women in Sericulture'.
UNIFEM (1988) 'What is UNIFEM,' New York: United Nations Publications (October).
UNIFEM (1989) Letter to Maria Spada.
UNIFEM (1990) 'UNIFEM/ITDG Technology Experience in Peru, Meeting Notes'.
UNIFEM (1991) *The World's Women; Trends and Statistics, 1970–1990*. New York: United Nations Publications.
UNIFEM (1991b) 'Peru Project Revision,' (March 22).
UNIFEM (1991c) 'Country Report: Peru.'
UNIFEM (1993) *UNIFEM Annual Report 1993*. New York: United Nations.
UNIFEM (1993b) 'Final Report on Women's Appropriate Technologies,' (February).
United States Agency for International Development (USAID) (n.d.) 'Environmental Assessment: Southern Honduras. Washington: USAID.
United States Information Agency, *World Fact Book 1992*. Washington: USIA, 1992.
Vargas, Virginia (1991) 'The Women's Movement in Peru; Rebellion into Action,' in Neuma Agular and Thais Corral, eds. *Development Alternatives for a New Era*. Rio de Janeiro: Editora Rosa dos Tempos.

Vellenga, Dorothy Dee (1986) 'Matriliny, Patriliny and Class Formation among Cocoa Farmers in Two Rural Areas of Ghana,' Claire Robertson and Iris Berger, eds., *Women and Class in Africa*. New York: Africana Publishing House.

Vengroff, Richard and Lucy Creevey (1994) 'Evaluation of Project Impact; ACEP Component of of the Community Enterprise Development Project,' Report to USAID (March 31).

Vorhips, M. (n.d.) 'Results of Mid Term Evaluation', ATI documents on Guatemala, Washington, ATI.

Whitcombe, R. and M. Carr (1982) 'Appropriate Technology Institutions; A Review,' *ITDG Occasional Papers* #7 London: Intermediate Technology Publications.

Wiley, Liz (1985) 'Tanzania: The Arusha Planning and Village Development Project,' in Catherine Overholt, Mary Anderson, Kathleen Cloud and James Austin, eds., *Gender Roles in Development Projects*, West Hartford: Kumarian Press.

Willoughby, Kevin W. (1990) *Technology Choice: A Critique of the Appropriate Technology Movement.* Boulder: Westview Press.

Wils, Frits (1993) 'Impact Study AWARE; Social and Political Aspects and General Evaluation,' The Hague: Institute of Social Studies (June).

World Bank (1992) *World Development Report 1992*. New York: Oxford University Press.

World Bank (1994) *World Development Report 1994*. New York: Oxford University Press.

Yudelman, Sally (1987) *Hopeful Openings; A Study of Five Women's Development Organizations in Latin America and the Carribean.* West Hartford: Kumarian Press.